Praise for *San Diego Bay: A Call for Conservation*

"An astounding feat of intellectual and moral synthesis. If there is a better, more future-thinking high school program anywhere in our galaxy, I haven't heard of it!"—**Carl Safina**, President, Blue Ocean Institute, School of Marine and Atmospheric Sciences, Stony Brook University

"This fourth book on San Diego Bay by the students of Gary and Jerri-Ann Jacobs High Tech High is a worthy follow-up to its highly acclaimed predecessors. This inspiring work takes an even more in-depth look at the astounding range of biodiversity that makes up San Diego Bay, and the intricate conservation challenges facing this vulnerable ecosystem. Through careful research and a series of thoughtful interviews, students explore causes of decline and propose solutions for the recovery of a number of threatened and endangered species. Part history, part politics, part biology, part artistic reflection, and all heart, this compilation is a remarkable testament to the power of project-based learning in a complex and changing world."
—**Allison Alberts**, Director of Conservation and Research, Zoological Society of San Diego

"*San Diego Bay: A Call for Conservation* examines the habitats and fauna of a region from a broad perspective of change through interaction with human impact. It is readable, diverse, thoughtful and informative in providing historical background, and a wide range of problems and approaches to their solutions. While I can highly recommend the content of the book to a broad audience it is the process of producing the book that is of scintillating brilliance. Natural history is asking questions of nature and using all the tools of science to seek answers. The faculty of High Tech High have led their students on a quest for natural history knowledge and how to fold this into an approach to sustainability. The greatest rewards for this effort will appear in the coming decades when these students participate in their communities as uniquely informed citizens who can help lead our civilization to a better future."—**Chuck Baxter**, Senior Lecturer, *emeritus*, Hopkins Marine Station, Stanford University

SAN DIEGO BAY
A Call for Conservation

by the students of
Gary and Jerri-Ann Jacobs High Tech High

Designed and Edited by
Sean Curtice, Megan Morikawa and Jennifer Zarzoso

This book was supported by the Unified Port of San Diego and the
National Sea Grant College Program of the
U.S. Department of Commerce's
National Oceanic and Atmospheric Administration under
NOAA Grant #NA08OAR4170669, project number C/P-1,
through the California Sea Grant College Program.
The views expressed herein do not necessarily reflect
the views of any of those organizations.

Published by California Sea Grant College Program
University of California, San Diego, California 92093-0232
Publication No. E-007
ISBN 978-1-888691-20-7

Dedicated to the Chula Vista Nature Center and its Staff

*For your commitment to the education of the
San Diego Bay community, and for your constant devotion
to the ecology and wildlife of the region.
We look forward to our continued work
together and our shared mission of conservation
and environmental awareness.*

Table of Contents

Predator-prey along an urban bay by Kelsey Hoffman, winner of the
Kurt Vavra Nature Photo Contest, 2007.

Foreword

If all politics is local, so ultimately is conservation. This superb compendium by the students of the Gary and Jerri-Ann Jacobs High Tech High is at every level a testament to that proposition.

San Diego Bay: A Call for Conservation has the advantage of addressing one of the most beautiful and biologically rich environments in America. From the nearby desert and river to shoreline, bay and ocean, its diverse fauna and flora are the self-supporting matrix within which the people of San Diego enjoy a high quality of life. Each of the species celebrated in these essays is worthy of a book by itself. Each is ancient—tens of thousands to millions of years old—and exquisitely well adapted to the environment of the California and northern Mexico coasts.

This collection of natural history and environment essays has a special importance for science education. It illustrates one point of entry to modern biology that now is opening wide. To an increasing degree, environmental biology and biodiversity studies are joining molecular and cell biology in the most important ranks of science. There is no better introduction to them than hands-on field studies, combined with theoretical studies and data analysis.

The celebration and close study of individual species in their natural habitat is also the best way to achieve conservation. People may understand the urgency of preserving the Amazon rainforest or the life of Asian rivers, but nothing approaches the motivation for saving life forms with which you are personally familiar. Their intimately enjoyed beauty and perceived value make them treasures we will not allow to be destroyed.

Edward O. Wilson, Pelegrino University Research Professor, Emeritus
Founder—E.O. Wilson Biodiversity Foundation

Preface

the Jane Goodall Institute

One of the problems with education in so many countries is a lack of freedom for teachers to design learning experiences that they know will benefit their students. High Tech High is unusual. Teachers are not only permitted but encouraged to think out of the box and they, in turn, involve their students in choosing, planning and executing projects that will enhance their learning experience.

We are at a crossroads—we humans have inflicted grievous harm on our environment bringing species to the brink of extinction, and entire ecosystems to the point of collapse. It is the youth of today that will suffer if, together, we cannot work to heal some of the damage: surely they deserve the opportunity to have a say in how this should be done? The students taking part in the projects described in *San Diego Bay: A Call for Conservation* are the scientists, engineers and policymakers of the future. And, too, they are the next generation of teachers and parents.

I have admired the farsighted policies of High Tech High ever since I met Jay Vavra. He introduced his students to our Roots & Shoots program: we share the similar goal of teaching youth about the problems we face and empowering them to take action to address those problems, to make choices that will make the world a better place.

Only recently has the importance of preserving biodiversity received the attention it deserves from politicians and the general public. And even when people do know, only too often they do little or nothing to help make a difference. Thus it is immensely gratifying to see that the students who have researched, written and designed this beautiful book address this threat, along with other problems such as pollution, development, invasive species and climate change.

One of the key elements in our Roots & Shoots program is to encourage members to pay special attention to *local* problems. Young people who study the ecology in their own "backyards," learn to understand and appreciate it. The High Tech High students have taken steps to ensure that their environment is preserved for the benefit and enjoyment of future generations. Moreover, they have learned the importance of acquiring solid facts before speaking out against actions that have harmed or are harming their environment: they are then able to articulate and explain their concerns and this approach can then lead to changes in the way people think and, sometimes, to changes in their behavior.

These students have become experts in their respective fields. They have learned the facts through their own research, rather than accepting all that they hear. It is firsthand information they share here. They are convincing and this book provides policymakers with pertinent facts that can be used to evaluate, for example, the current status of local endangered species, the factors that contribute to these threats, and clear suggestions about how these should be tackled.

I know full well the amount of work and dedication that has been necessary to carry out the diverse range of original research presented by these students. Their investigations have included population surveys, behavioral observation, and the recovery and release of light-footed clapper rails, to the sophisticated molecular biology of DNA barcoding. Collectively their work comprises a scientifically meticulous and incredibly sophisticated study.

Most people are amazed that such important results can be achieved by high school students. I am not one of those! I know this group of students—the "High Tech High Roots & Shoots" team—and have become familiar with their high standard of excellence. Each year they design innovative ways of working within their communities to improve the lives not only of wildlife but also the human population of San Diego.

One of my reasons for hope is the growing number of young people who care about the future of Planet Earth. These students exemplify the kind of informed, passionate, determined and active young people who can, together, create a more harmonious future, which is immensely reassuring since they are the future stewards.

Jane Goodall Ph.D., DBE
Founder—the Jane Goodall Institute and U.N. Messenger of Peace
www.janegoodall.org
www.rootsandshoots.org

Acknowledgments

The intertidal habitat along the shore of San Diego Bay is only complete if a diverse set of individuals act in concert, coexisting to form a thing of natural beauty. Such is also the case with the composition and construction of this book. In the following pages, we recognize the importance of individuals and organizations whom we have come to call "Stewards of the Bay" and who seem to have keystone roles within the region.

The students who contributed to this book first extend thanks to the teachers who organized the project: Dr. Jay Vavra and Mr. Tom Fehrenbacher. There are few educators in this world who hold such high expectations for high school students, and subsequently inspire students to meet these expectations.

During the course of this book's production, we were able to make contact with Dr. Edward O. Wilson, leading myrmecologist and, as Dr. Vavra has said, "the greatest biologist alive today." We are privileged to have such an influential scientist take the time to lend his voice to our book by writing a truly remarkable foreword, and extend our deepest and sincerest thanks to him.

Dr. Jane Goodall, world-renowned biologist and humanitarian, has been a friend and inspiration of our Roots & Shoots program for several years now. Dr. Goodall has been an incredible inspiration to all of us who have worked on this project. We are extremely grateful to her not only for her help with our San Diego Bay series, but for the work she has done to better the entire world.

We thank Dr. Exequiel Ezcurra, former provost of the San Diego Natural History Museum, for sharing his extensive knowledge of biodiversity with us. His interview is an extremely valuable addition to the book, and we look forward to our continued collaboration with him.

On behalf of ourselves and the wildlife of the region, we thank Jim Peugh, wetlands expert and conservation director of the San Diego Audubon Society, for his efforts to restore and protect local wetlands. We also greatly appreciate his perspective on wetlands and editorial comments for this book.

Dr. Jeremy Jackson of Scripps Institution of Oceanography and Shiftingbaselines.org was a great help in addressing and explaining the complex and delicate issue that is shifting baselines. We would also like to thank Dr. Randy Olson, cofounder of Shiftingbaselines.org for his humor, help and insights on the topic. Dr. Olson's approach to creatively mainstreaming conservation awareness was also inspiring.

The DNA barcoding work was supported by the collaboration of Dr. Oliver Ryder and his amazing laboratory: researchers Leona Chemnick, Heidi Davis, Jennifer Lau and Maggie Reinbold in the Genetics Department, Division of Conservation and Research for Endangered Species of the San Diego Zoological Society. We would like to thank Invitrogen Corporation for supplying the kits and materials to perform the molecular techniques in High Tech High's laboratory. Special thanks also go to Invitrogen for the opportunity to present our research at the BIO 2008 meeting.

In telling the story of the green sea turtle in the bay, we thank assistant professor Margie Stinson of Southwestern College for her first-hand account of Carl Hubbs. Thanks to Peter Dutton, National Marine Fisheries Service—one of the world's experts on sea turtles—who also helped us tell the unique story of the bay's turtles. We appreciate Ed Parnell of the Integrative Oceanography Division, Scripps Institution of Oceanography, for his review of this chapter.

Thanks to Richard Gilb, environmental affairs manager with the San Diego County Regional Airport Authority, and his staff for assisting us in observation and documentation of the unique California least tern nesting situation at the airport. We also appreciate the efforts and information of Chris Redfern of the San Diego Audubon Society, on terns and local conservation issues.

Charles Gailband, curator of animals at the Chula Vista Nature Center, has been one of our most involved collaborators throughout this project. His constant willingness to lend his help and his extensive knowledge on the light-footed clapper rail, as well as nearly every other species presented in this book, were an enormous asset to us, and we are immensely grateful to him. Many thanks to Dr. Richard Zembal, noted clapper rail expert, who allowed student researchers to attend numerous surveys, releases and other events on the species. He also provided helpful information and data on this charming and important bird.

Thanks to Will Sooter, naturalist, photographer and peregrine falcon enthusiast, who took us into his "office" along the majestic cliffs of Torrey Pines and shared the awe-inspiring beauty and prowess of falcons.

We encountered and observed many of the book's avian protagonists with the help of Brian Collins, U.S. Fish and Wildlife Service refuge manager who showed us amazing nesting habitats at the South Bay Salt Works and the mouth of the Tijuana Estuary. Jeff Lincer aided in compiling information on the effects of DDT on species covered in our study. We would also like to thank him for his extensive knowledge on the topic and his review of several chapters dealing with raptors. Robert Patton shared his years of experience observing coastal birds and also allowed us to assist him in banding studies and nest surveys. We relied upon Mr. Patton's expertise for proper identification of photographs taken of suspected tern species around San Diego Bay. Both Mr. Collins and Mr. Patton were especially helpful with editorial comments on many bird species in this book.

Thanks to Maris Sidenstecker, founder of Save the Whales, who shared her life story that is so closely tied to the whales of the world. We offer our hope that her work to preserve whales will continue to benefit some of Earth's rarest and most threatened marine mammals. Appreciation goes to Wayne Perryman, NOAA/Southwest Fisheries Science Center for his thoughtful comments on the gray whale.

Many thanks go to David Merk at the Port of San Diego. He was able to provide important insights into the issues that ebb and flow around ship traffic in San Diego Bay and their associated effects on the area's urban environment.

We appreciate the help of Jamie Gonzalez. Her interview successfully introduced us to the seriousness of invasive species in San Diego Bay and around the world.

Special thanks go to Mitch Perdue of the U.S. Navy. After giving students an exclusive tour of San Diego Bay by boat, he treated us to an *in situ* interview on an island created from dredged spoils.

We appreciate Bruce Reznik of San Diego Coastkeeper for providing a series of interviews for this book. Both the area's wildlife and this book's authors sincerely thank him for his life's efforts to protecting the environment.

We had the honor of meeting and interviewing Nobel Prize winner Dr. Michael Oppenheimer, a leading researcher on the Intergovernmental Panel on Climate Change. We thank Dr. Oppenheimer for his support, and for his important international work to reverse global warming. We also thank Dr. Tony Haymet, director of Scripps Institution of Oceanography, both for his help and for his dedication in alerting the public to the dangers of climate change.

Dr. Brent Stewart was instrumental in the creation of our piece on the Baiji River dolphin, a eulogy in the making. We thank him for providing crucial information on the dolphin and for graphically illustrating the tragedy of a species lost forever.

We greatly appreciate the efforts of Dr. Brian Joseph, John Lopez, Ben Vallejos and Dan Beintema of the Chula Vista Nature Center, who not only endeavored to preserve the Sweetwater rainbow trout, but also provided suggestions and made contributions to our story on the trout's delicate situation. The Nature Center itself is now on the brink of extinction, and we continue to fight to save this invaluable community resource.

Our correspondence with Rachel Muir, imperiled species coordinator of the U.S. Geological Survey, was another experience for which we are grateful. The insight into imperiled species, the information on the Endangered Species Act, and the prestige of the Muir legacy have benefitted our book immensely. We also thank Dr. Muir for the collaboration formed with Dr. Vavra in chairing a session on biodiversity education at the ninth annual meeting of the National Conference for Science, Policy and the Environment. They made us proud by presenting this book as a model for biodiversity education in the Unites States.

We would like to thank Dr. Kevin Hovel of San Diego State University for revealing the importance of eelgrass restoration and preservation. His interview shed light on the murky waters of the issue, allowing inspiration and providing a fresh, green habitat for our ideas.

Jack Webster, sustainable fisherman and chair of the American Albacore Fishing Association, helped us decipher the difference between sustainable and unsustainable fishing practices, and allowed us to understand the importance and necessity of the former. Thanks also to Dr. Heidi Dewar of NOAA/Southwest Fisheries Science Center for giving us an important scientific perspective on issues dealing with sustainable fishing.

We would like thank Andrea Compton, chief of Natural Resource Science at the Cabrillo National Monument for her insights on the importance of marine protected areas. Special thanks also to Dr. Paul Dayton, Integrative Oceanography Division, Scripps Institution of Oceanography, for his editorial comments on marine protected areas.

Deborah Day of Scripps Institution of Oceanography reviewed the life story of the amazing Dr. Carl Hubbs and we thank her for her comments. Hubbs's son, Dr. Clark Hubbs, professor emeritus at the University of Texas, was another contact that we are honored to have made. Sadly, Dr. Clark Hubbs passed away just months after we interviewed him and never got to see this book. However, the conservation spirit of father and son live on and we hope their successful scientific careers will inspire countless others.

Jon Christensen, writer, historian and Steinbeck Fellow, suggested a number of resources that were of great help in the examination of environmental holism. He graciously provided feedback to our summation. We also appreciate the insight into holism and the life of Ed Ricketts from his contemporary, Dr. Charles Baxter of Stanford University's Hopkins Marine Station.

A number of the outstanding student photographs found in this book were submissions to the Kurt Vavra Nature Photo Contest held at High Tech High. Special thanks to Larry and Nancy Kuntz of Nelson Photo for sponsoring this contest and motivating many talented individuals to provide the images that are now included in this book. Thanks to Robert Vavra for supporting and judging the photo contest as well.

Thanks to Ann Bowles, senior researcher of Hubbs-SeaWorld Research Institute, for being a part of this project from beginning to end. We appreciate her advice on the scientific approach, her willingness to lend a critical ear, and the new leads she provided for our stories.

We appreciate the biological wisdom provided by Richard "Pancho" Lantz, philosopher, poet and former biology professor, during the planning stages of this project and also during our field work.

There are some individuals who, despite no direct obligation to this book, volunteered their time to edit sections nonetheless, and therefore deserve special recognition. We would like to thank Judith Shushan, Gale Vavra, Ron Vavra, former High Tech High student editors Natalie Linton and Bryndan Bedel, Dr. Ron Burton (Scripps Institution of Oceanography), Rick Stallcup and Melissa Pitkin (PRBO Conservation Science) and other anonymous contributors.

Thanks to Kelly Makley, the environmental education director for the Port of San Diego, for her vision and unending support and advice throughout the project. We also greatly appreciated her helpful suggestions when editing the final drafts.

We greatly appreciate the advice and editing expertise of Joann Furse and Marsha Gear of the Communications Department of California Sea Grant. Their diligence and belief in this project brought it into book form. We also thank Rene Carmichael for editing a number of the chapters.

Thanks to High Tech High and the Regional Occupation Program for supporting this project through and through and being champions of innovative project-based learning.

Teacher Introduction

Our concern for the environment has grown over the years. As teachers in biology, humanities and mathematics, we have found, together with our students, a gradual deterioration in our natural surroundings. The study of our local environment, San Diego Bay, has told us that the environment left to us today is nothing like that originally witnessed by the bay's early humans. We know, with our student authors, that we have inherited an environment much diminished in diversity and sustainability.

From previous generations our students have acquired a disease not of their making. Those coming of age today face a planet made sick by climate change, species extinction, human population growth and shortsighted human interaction with the environment.

And, we find these changes have occurred largely unnoticed. Using biology's concept of a "shifting baseline," we know that this much-diminished landscape, a long time in the making, will only continue its decline if we fail to first recognize the problem and take steps to address it.

This book is the fourth in a series written and produced by our students. Through the books, we hoped to empower our students with meaningful work and help provide them with a greater appreciation for nature. *San Diego Bay: A Call for Conservation,* examines the bay's current state of health, discusses reasons for its decline, and provides positive solutions. It is our hope it will inform, alert and help other concerned individuals and communities with their own conservation and restoration efforts.

In the following pages, student authors express their findings and concern through original research and original commentary. The chapters address their topics through a review of literature, fieldwork, DNA barcoding, the taxonomy of organisms and interviews with local experts. Original poetry, nature reflections and photography provide insight, commentary and student perspective. While, as teachers, we planned the book's chapter headings and general outline as much as possible, along with the topics to be covered and research to be done, our year together also took its own twists and turns.

Some of our story unfolded before us and made teachable moments that could not have been scripted. The killdeer nesting in the school parking lot over spring break showed us with heartbreaking relevance what the loss of habitat means to precious killdeer chicks. We learned first hand that we can help distraught wildlife. Upon further reflection, we found, even in this misplaced nest, a message of hope; when creatures adapt to hostile environments, they are reclaiming part of what was lost.

In another unscripted event, attending Brent Stewart's talk on the now extinct Baiji Yangtze River dolphin was both poignant and prescient. His tragic tale revealed the universal warning in much of this book, should we choose to listen. We cannot recover an extinct species, just as the bay's environment may not recover from thoughtless land use, ship traffic and pollution. Together, with our students, we have come to the conclusion that if we ignore our environment's baseline, diversity or natural refuges, we do so at our own peril.

While much of contemporary society focuses upon the role of the individual, we found hope in the work of gifted individuals and in collective efforts. In their final chapters, our students considered the far-reaching work of Carl Hubbs, San Diego's own great marine biologist, and the many different collective agencies known as the Stewards of the Bay. Our authors found in these local agencies and nonprofit organizations a collective effort and common mission to help heal the bay. As part of our own contributions, we formed new alliances with local environmental concerns and agencies. Bringing together research partners at the San Diego Zoological Society and innovative products and support of the biotech industry from Invitrogen Corporation, students were able to apply novel molecular tools to describe important components of this conservation story.

Edward O. Wilson has diagnosed all of us with a case of Biophilia. In this positive conclusion, Wilson sees each of us with an innate sense of affection for life. Has humanity somehow made itself immune from this wonderful condition? We answer "no." Though, at times, we act as if we are separate or apart from our natural surroundings, we find hope in Wilson's belief that we have a compassion for nature.

Our students arrived in class with a feeling of disconnection from their natural world. By addressing this, the authors rekindled their primal connection and renewed a compassion for life and the environment. In order to conduct the study, complete the research and finish the writing, the truth of our integral place in nature reemerged. The project itself shows the spirit of these students and those community experts who assisted in their explorations. Through fieldtrips, class seminars, naturalist literature, DNA technology, statistics, labs and work in natural history, we concluded that a simple shift in consciousness will allow us to see we are inextricably part of one earth, one environment. As such, we find ourselves called to conservation.

Jay Vavra, biology teacher
Tom Fehrenbacher, humanities teacher

Gary and Jerri-Ann Jacobs High Tech High

Student Introduction

It is hard to believe that in less than six years, students from High Tech High have produced and published three books detailing the beauty and history of San Diego Bay. *Two Sides of the Boat Channel*, *Perspectives* and *San Diego Bay: A Story of Exploitation and Restoration* are the publicly acclaimed precursors to this story, which now looks into the unique history of conservation in the San Diego Bay area. *San Diego Bay: A Call for Conservation* is an extensive analysis of the delicate balance between humans and nature in an urbanized ecosystem. Through this analysis, we attempt to promote a need for conservation in this diverse ecosystem we all call home.

In our studies, we discovered that San Diego is home to an array of unique species, due mostly to the wide range of habitat the bay has to offer. This delicate biodiversity, as acclaimed biologist and author of this book's foreword Dr. E. O. Wilson emphasizes, is the key to a rich ecosystem. We attempt to demonstrate the incredible amount of diversity that exists in our own backyards.

Additionally, the land in San Diego also creates a diversity of ecosystems. The wetlands that surround and weave through the city limits are home to many wildlife species. With the incredible conservation efforts of local organizations, these wetlands serve as a prime story of a constant battle for restoration and conservation. They are the habitats that we must fight to protect in order to save the homes of these species.

As we observe the change in these ecosystems over time, we realize our dependence on "perspectives" to determine whether or not the habitats are truly natural. This dependence on perspective to define "nature" is represented in an ecological buzz phrase called shifting baselines, a concept heavily studied by professors at the University of California in San Diego. Through this concept, we hope to portray the flux of nature and our innate inability to fully understand what is truly natural in the bay.

These themes are kept throughout the book as we began to understand the threatened species of the bay. The book includes a short section on methods used to study and assess information for our story. We then focus our attention on the species, mostly avian, which are threatened by human activity. An outline of each organism's biology is given at the start of their section. Students wanted to focus on the unique living habits, dietary patterns, and morphological features that were unique to these species. Then, each species is analyzed for the cause of its loss from the bay. From pesticide to habitat destruction, the next section attempts to pinpoint the causative agents, which have led to population decline.

Finally, the recovery section attempts to highlight the successful efforts, to bring these beloved organisms back, often from the brink of extinction. Unfortunately, these species are often only afforded the protection that our society's resources allow. We hope to demonstrate the need for continual effort in conserving the unique species that make up our bay, and, equally importantly, preserving their habitats. The species were chosen for their degree of imperilment or lack of recognition in terms of conservation efforts. Our list of species represents only a small portion of those affected, both positively and negatively, by man's presence in San Diego.

In the next section of our story, we pay close attention to the activities and ecological events that have directly or indirectly contributed to the need for conservation. As one of the of the major bays on the West Coast, San Diego Bay has a heavy amount of traffic from commercial, military and recreational vessels, often traveling to and from foreign waters. Student researchers highlight the pollution, exploitation and environmentally unhealthy urban growth in San Diego. Another unnatural cause of stress on the environment comes from the invasive species introduced by maritime activity. Invasives such as the Asian mussel or a type of tunicate threaten San Diego Bay by throwing the delicate ecosystem off balance. This section highlights those and other creatures that are having a detrimental effect on the bay's endemic species.

Yet one of the largest human impacts on the bay has been the coastal development and reconstruction that has occurred in this beach city. The effects of dredging as well as increasing land use have destroyed crucial habitats and pushed creatures to the test. Together, these pieces highlight the constant battle of man versus nature in the form of land development. Yet one of the most widespread and unavoidable effects has sprung from the global changes in our climate. Thus, our final topic of the section focuses on the effects of climate change worldwide and how that relates locally to San Diego.

In spite of the list of causes for decline in species around San Diego, this city is also an incredible resource for restoration and conservation efforts. The final section of the book deals with the fight for the survival of species locally and globally. The first solution focuses on the national Endangered Species Act and how that has affected local efforts through raised awareness, federal protection of species, and increased regulation enforcement. In addition to national effort, there are local stewards of our bay, the individuals and organizations that donate time and effort to the survival and sustainability of our diverse ecosystem, its endangered creatures and habitats. We thank our stewards for keeping the message of conservation and hope alive in San Diego restoration efforts. One specific effort of interest is the eelgrass restoration project in San Diego's coastal waters. We felt this to be a staple story of detriment that led to concern that led to action. Eelgrass restoration shows how our efforts can truly make a difference. Another effort focuses on the sustainable fishing practices, a global concept adopted in San Diego. Sustainable fisheries strive to practice environmentally friendly methods to manage their catches, as well as promote a solution to overfishing. The concept demonstrates how communities can work together to save species while still maintaining effective levels of production. The land itself has proved to be a ground for great strides in protection. Yet San Diego is unique in its coastal extension to the heavily trafficked seas and we focus on the marine protected areas that protect our beautiful oceans.

Finally, we focus on the individuals who greatly contributed to local restoration efforts. One man in particular, Carl Leavitt Hubbs, is notable by expanding our knowledge of the natural world and changing the way we view the environment. We hope to commemorate his efforts by demonstrating just how much of an impact Carl Hubbs, a true student of holistic environmental philosophy had in San Diego.

These organisms, problems and solutions all come together to create an extensive overview of conservation needs in San Diego. We are not only high school students, but students of the school of environmental holism, following in the wake of John Steinbeck and Ed Ricketts, who have taught us the importance of applying connection, complexity and compassion in order to sustain life. Our ongoing studies strive to demonstrate the importance of environmental awareness and action, and to show that individuals can make a profound difference. Thus, our book is not just a story, but rather a call for conservation that we hope will echo across the beautiful city of San Diego and beyond, for the benefit of our biosphere and all its inhabitants.

Gary and Jerri-Ann Jacobs High Tech High Student Authors

Biodiversity

As the world's natural resources dwindle and habitats become more affected by human contact, biodiversity suffers. Biodiversity, in the simplest of terms, encompasses all life and is defined as the variety of organisms within a given ecosystem. The balance of biodiversity is so complex that it is often not fully understood, so significant that organisms miles apart depend on one another for survival, and so delicate that the absence of any one species can break it. Organisms are interconnected by the role that they play in balancing an ecosystem, for example: producers create food (and often oxygen) for other organisms. Primary consumers control the producer population and secondary consumers control the primary consumer population, and so on. Decomposers convert decaying matter into energy so that the process can continue. Biological relationships can be viewed on a scale as large as the entire Earth or as small as a community within another organism (seen with endosymbiosis, one organism living within another). Within the past decade, extensive research has made the United States one of the most biologically studied and preserved nations in the world. (Stein, Kutner, and Adams 2000)

The United States contains the California Floristic Province, one of just 34 biological "hotspots" in the entire world, or regions declared to be of invaluable biological significance due to their biodiverse ecosystems and endemic species. San Diego County is an extremely varied region of the wholly diverse Floristic Province that encompasses most of the state and all of its coastline. The county is home to a wide variety of native flora and fauna, and has thus received a great deal of attention from environmental and ecological scientific communities. In addition, diverse coastal terrain with

"We have no given right to get rid of other species on Earth.... We have a moral responsibility to maintain the flow of life..."

— Exequiel Ezcurra

rocky open coast, estuarine, bay, mudflat and salt marsh habitats, combined with a unique Mediterranean climate, make it one of the most significant avian breeding areas in the world. ("California Floristic Province" 2007)

Despite the existence of relatively biodiverse regions, the whole of nature's dselicate balance continues to be threatened by human interference. Deforestation, dredging, draining of wetlands, pollution and poaching have caused the extinction of countless species and have impacted the planet at an alarming rate, especially in recent years. While progress has been made towards restoring the wild populations of many threatened species, few populations have returned to their pre-impact numbers. (Groombridge and Jenkins 2002)

The most common and basic means of measuring biodiversity is "species richness," or the number of species present in a given area. Scientists have estimated that the Earth as a whole is home to anywhere from 3–100 million species of all kinds, with 14 million being the most commonly accepted number. It is estimated that over 250,000 species are native to the United States. Broken down, the species composition of the nation includes 34,000 of the world's 56,200 species of fungi, 18,100 of the 265,000 species of plants, and 148,000 of the 1,074,000 known animals. (Stein, Kutner, and Adams 2000)

The variety of biodiversity may be defined very simply through creatures with entirely unique adaptations or characteristics. Many such organisms exist in the United States. Entire groves of certain aspen trees (*Populus tremuloides*) are technically a single organism; each tree is connected to the next trunk through a common root system; they are all genetically identical. In Utah, one of these forests that is actually one individual tree containing more than 42,000 trunks and covering 107 acres of land. Many species known as "living fossils" exist today, including the sturgeon, horseshoe crab, and coelacanth. Various species of sturgeon (*Acipenser* and *Scaphirhynchus*), armor plated fish that are often over six meters in length, have remained unchanged since the Devonian period, or "Age of Fishes," for 400 million years. The horseshoe crab (*Limulus polyphemus*) is one of four members of an animal group from the Cambrian period of more than 500 million years ago. The coelacanth (genus *Latimeria*) appears in fossil records dated 400 million years ago. (Stein, Kutner, and Adams 2000)

Even more so than the United States as a whole, California is internationally recognized as a global hotspot for biodiversity. The California Floristic Province is both one of the most diverse and most threatened areas in the world. It is home to the coast redwood (*Sequoia sempervirens*), the tallest tree in the world at 111 meters high, and the daunting giant sequoia (*Sequoiadendron giganteum*), the largest tree in the world, with trunks that can exceed six meters in diameter. California is well known for its tremendous biodiversity and endemism, the phenomenon of a species being native to only one area in the world and playing an important role in gauging a region's biologic condition. In addition to being unique, or rather because they are unique, endemic species are some of the most fragile anywhere—harming just one region can lead to the extinction of an entire species. The giant sequoia and coast redwood are both endemic species widely known for being biological extremes native only to California. Including these two giants, there are an incredible 2,124 endemic plant species in the Floristic Province. In addition to its incredibly vast plant life, the Province is the largest avian breeding ground in the United States. ("California Floristic Province" 2007) (Stein, Kutner, and Adams 2000)

San Diego County, within the southern-most reaches of the Floristic Province, is home to a wide variety of flora and fauna. This is largely due to the temperate Mediterranean climate of the

The varied geography of coastal San Diego provides biodiverse habitats. Left to right: wetland, mudflat, coastal bluff, rocky headland.

California coastline and the diverse terrain present in the large county. Sandy beaches, salt marshes, deciduous forests, deserts, riparian woodland, coastal sagebrush and evergreen forests, among other smaller coastal ecosystems, are all found within the county limits. This varying terrain is accompanied by a temperate climate in most local ecosystems, but both significantly colder and hotter environments can be found in mountainous or desert terrain. Most notably, the nation is one of the best areas in the country for avian breeding, so much so that it has been designated a Globally Important Bird Area by the American Bird Conservancy. In addition to the unique sagebrush scrub habitats, the vernal pools are another ecosystem found here.

San Diego County

Pacific Ocean

■ Biodiversity "Hotspots"

The California Floristic Province.
Data compliments of Biodiversityhotspots.org

These isolated, stagnant pools are an especially rare type of wetland habitat that collect rainfall and where, for a short while in spring, an abundance of blooming flowers and rare animals can be found. The vernal pools of San Diego have survived somewhere between 125,000 and 400,000 years and were relatively untouched until the 1980s when the human population greatly expanded in the region. Today, only 3% of the county's vernal pools remain, and most are protected in the San Diego National Wildlife Refuge Complex. (City of San Diego Vernal Pool Inventory 2004)

Countless ecosystems throughout the nation are threatened by widespread development. Specifically, native species in California and San Diego County are affected by widespread coastal development. Only the results of climate change discussed later in this book may result in the "reclamation" of this habitat.

Commercial farming is also a serious threat to natural habitats within the state, especially considering the nation's growing population and the resulting increased demand for crops. In fact, over one-half of the nation's agricultural products are grown in California. The ecological footprint of commercial farming is large when aspects of runoff water carrying pesticides and fertilizer are considered.

The California hotspot for biodiversity originally stretched 182,000 square miles, but today only 45,600 square miles of natural vegetation remain, or some 24.7% of the pristine state.

The populations of 18 vertebrate species endemic to the Floristic Province are threatened. Besides the marine mammals, sea turtle, sea birds and raptors described later, included are the critically endangered island gray fox (*Urocyon littoralis*), giant kangaroo rat (*Dipodomys ingens*), Guadalupe storm petrel (*Oceanodroma macrodactyla*), and California condor (*Gymnogyps californianus*). Though typically not as discussed as plants or vertebrates, an incredible level of invertebrate diversity exists within the province. In addition to the abalone described later and many other marine species, about 28,000 species of insects, or roughly 30% of all known insects in the United States and Canada, are found in California. Of these, 9,000 are endemic and therefore especially vulnerable to extinction. The reach of environmental degradation is not limited to larger animals and each organism, no matter how small or poorly-understood, plays a vital role in the biosphere. ("California Floristic Province" 2007)

Some of the most significant threats to natural biodiversity are discussed further in later sections of this book. These include, but are not limited to: invasive species, ship traffic, land development, coastal dredging and climate change. In our modern word where travel is limitless and development widespread, it is very easy for an alien organism to be introduced. Native species, due to natural selection and their adaptation to local climates, form a complete biological web with a fluid balance. Invasive species disrupt that balance. Many of these plants are fast growing and can crowd indigenous species until competition becomes too great for them. Introduced animals cause considerable damage to natural plant life. While the actual number is not known, it is estimated that 3,500 nonindigenous plant species and 2,300 non-native animals now inhabit the United States. Despite their limited numbers, introduced plants and animals devastate native

biodiversity because they are often free of any natural predators that served as population control in their former, indigenous habitats. A local example occurred in 2000 when an outbreak of the seaweed *Caulerpa taxifolia* that has devastated the northern Mediterranean Sea threatened a local coastline before its eventual containment. ("National Invasive Species Center" 2007)

Several international agreements aimed at preserving biodiversity by conserving threatened species and their delicate habitats have emerged since the 1970s. One of the most significant came in 1992 when the United Nations Conference on Environment and Development, also known as the Earth Summit, was held in Rio de Janeiro. This pivotal summit resulted in an agreement on the "Convention on Biological Diversity," a plan that revolved around conservation and would work towards sustainable development on an international level. Despite the efforts of some and even the successful restoration of select species and habitats, the overall environmental situation has worsened since the summit due to the exponential increase of carbon dioxide in the atmosphere and rapid population growth. (Groombridge and Jenkins 2002)

As it becomes more critical for survival, restoring ecosystems also becomes more difficult. Combating widespread human development is no simple task considering the rapid population growth that is occurring on a global level. Without drastic reform, it is likely that pressures on biodiversity will only escalate in the future. While the main issue in developed nations such as the United States is to preserve natural habitats and species, developing nations present a unique problem. Many of them still embrace a model of capitalism reliant on aggressive growth, one that cannot continue if the planet's environmental and ecological crises are to be resolved. Minimizing the consequential environmental impacts while combating extreme poverty will prove to be one of the greatest environmental challenges of our future. A nation that is losing its people to genocide and epidemics is even less likely to worry about a small, endemic bird nearing extinction. Densely populated regions of the world, such as China and India, will be especially difficult to restore. (Groombridge and Jenkins 2002)

A major solution working to protect the naturally rich areas of the world is the creation of refuges or protected areas. The Convention on Biological Diversity defines a protected area as "a geographically defined area, which is designated or regulated and managed to achieve specific conservation objectives." First established in the 1940s, the number of protected areas has increased from about 500 to more than 2,200 internationally over the last 30 years. Still, the ratio of protected to nonprotected areas remains low. Defining and establishing these protected areas is central to preserving some of the world's most endangered species that inhabit shrinking natural ecosystems. ("Protected Areas" 2007)

There are a number of factors used to determine which areas to protect, including endemism levels, the uniqueness of the habitat and species richness. Considering the significance of these parameters, it is no surprise that the Floristic Province is a "hotspot." The hotspot system, a tool principally developed by Conservation International, serves to identify areas of high species endemism and high degrees of habitat loss to direct conservation planning in the face of possible development. Altogether, there are 34 hotspots in the world that include the Philippines, the Atlantic forests of Brazil, Japan and the Himalayas, among others. An interesting study has been conducted to determine the most biologically significant regions on a larger level. It ranks all countries based on their estimated species richness and endemism, both in land vertebrates and in plants. Of the world's diverse waters, Marine Protected Areas (MPAs) are widely regarded as essential to maintaining aquatic biodiversity. Select over-exploited fisheries, such as those surrounding Japan, and sensitive coral reefs, including the Great Barrier Reef in Australia, are now MPAs. Yet, like their land-based counterparts, MPAs protect a total amount of water much smaller than the ideal. ("Protected Areas" 2007) ("California Floristic Province" 2007)

While protecting small, specific regions is generally beneficial, improving organization can create more interconnected preserves and curb many of the complicated legal issues associated with environmental conflicts. For decades, individual species were variously considered rare or endangered by the County of San Diego, the State of California, the U.S. government, the U.S. Fish and Wildlife Service and the California Department of Fish and Game. Varied reports meant scattered responses from government entities, private conservationists, wildlife agencies and developers that were all but consistent or organized. The County of San Diego, in collaboration with the federal government, conservationists, wildlife agencies and developers, first implemented its solution in 1998: the "Multiple Species Conservation Program" (MSCP). Rather than protecting small areas either separated from other preserved land or surrounded by large-scale urban development, the MSCP aims to create large interconnected preserves in more open land to benefit a number of species at the habitat level. (City of San Diego Vernal Pool Inventory 2004)

Select species continue to be protected through more tailored solutions, but the MSCP is at work to protect much of the undeveloped land in the county while it still exists. Free of expensive private costs, multiple permit authorities and negotiations on a project-by-project basis that plagued such efforts in the past, the MSCP outlined

three plans to create a more effective preserve system: the South County MSCP Subarea Plan, the North County Subarea Plan, and the East County Subarea Plan. Together they work to benefit some 300 sensitive, rare, threatened and endangered plant and animal species. Preservative and restorative efforts tailored to specific species under each plan are dictated by habitat maps, regular surveys, various reports and development plans. Although it is extremely vital to individual species, the MSCP's significance is best illustrated through acreage. It aims to acquire or permanently protect 172,000 acres, 98,379 of which are classified as undeveloped, of the total 582,000 acres of land examined. In years since the plan's approval, thousands of acres have been added to that goal by local, state and federal agencies. (City of San Diego Vernal Pool Inventory 2004)

While protected areas are extremely beneficial steps towards maintaining biodiversity, preserving our current ecological conditions may no longer be enough. With the rapidly increasing human population, preservation alone cannot combat such environmental degradation. An increasing awareness of the negative human impact has developed interest in environmental restoration. Environmental restoration is a crucial complement to protecting areas, and it is widely accepted that restoration will become the central practice for environmentalists of the future. The main goal of such restoration is to return ecosystems to their original structure and function, to their pristine state.

Programs aimed at reintroducing endemic species to their natural habitats will also greatly contribute to restorative efforts. (Groombridge and Jenkins 2002)

Preserving biodiversity becomes more possible as it is understood more deeply. Attempting to reveal all of nature's complex interactions is likely far too ambitious to be immediately worthwhile, but the compilation and analysis of global population data could be the first step in that direction. The Internet and recent advances in compact wireless technology have created a global platform from which to organize such data, and the "Encyclopedia of Life" (EOL) takes full advantage. Constantly growing, collaborative and truly international, the EOL is a combination of websites and data that makes unique information about species richness and diversity readily available to anyone, anywhere. Its goal is to create "a constantly evolving encyclopedia that lives on the Internet, with contributions from scientists and amateurs alike." Someone in the field will be able to upload pictures and observations from their cell phone while a person on another continent sitting at their computer will be able to access that information almost instantly. The brainchild of renowned conservationist E.O. Wilson, the EOL has tremendous potential made possible by advancements in technology and a global community that is becoming more conscious of the world it shares.

Edward Osborne (E.O.) Wilson, renowned naturalist, researcher and secular humanist, is one of the world's foremost experts on biological diversity. Born in Birmingham, Alabama, Wilson's interest in nature grew from an early age when he began surveying local ant populations. He received his bachelor's and master's degrees from the University of Alabama and a Ph.D. from Harvard University, where he teaches today as Pellegrino Research Professor in Entomology for the Department of Organismic and Evolutionary Biology. His many accomplishments include the formulation of sociobiology, "the systematic study of the biological basis of all social behavior." He established the new scientific field by applying evolutionary principles to understanding the social behavior of animals and humans. He examined consilience, a synthesis of knowledge from different fields, between the sciences and humanities through methods that have been used to unite scientific disciplines. In addition, Wilson studied the mass extinctions of the twentieth century and their ecological relationship to modern society. He has received more than 100 awards and honors including Pulitzer Prizes for *On Human Nature* and *The Ants*, and the National Medal of Science. That one individual can impact the environment so strongly implies limitless potential for the capabilities of society. ("Wilson Life and Work" 2008)

While various scientists and environmentalists recognize the significance of environmental restoration, government action and public awareness are essential for it to begin on a large scale. Various international conventions and treaties have contributed greatly towards environmental preservation, but individual governments have remained comparatively uninvolved in restoration due to the high cost. As has been true for global warming, experts fear that governments will only become involved when the environment has passed a point of no return. Due to the complexity of nature's balance, it is almost impossible to determine when such a threshold will be passed. As suggested by dramatically increasing global temperatures, it may have already. Although dramatic environmental initiatives have yet to become the norm at an international level, public influence is capable of nudging the developed world into action. Simply breathing oxygen connects every human to the trillions of organisms that produced it. It is becoming clear that our once strong connection to the natural world is breaking, and reversing that trend demands that individuals, organizations and governments work as one. (Groombridge and Jenkins 2002)

Action is being taken to lessen the impact of human development upon natural habitats and native species. Local, state, and federal government programs like the MSCP cannot independently accomplish such daunting ecological goals. Nonprofit environmental organizations and awareness groups play major roles in preservation and restoration both locally and nationally. Some of these to be discussed further include

High Tech High students assess the change in biodiversity by using transects to count the number of invertebrates present in an ecosystem.

the San Diego Audubon Society, Pro Peninsula, Friends of Famosa Slough, the National Wildlife Refuge Association and the Chula Vista Nature Center. Select individuals, including Jim Peugh of the Audubon Society, have introduced successful restorative projects already making an impact around the county. ("SDAS Conservation Projects") The San Diego Natural History Museum's "San Diego Bird Atlas" that compiles population data for the many avian species present throughout the county is heavily reliant upon volunteer observers. (Unitt 2004)

The Famosa Slough project is a model public potential and small-scale restoration. The wetlands have been revitalized and species that once inhabited them are starting to return because of the efforts of some environmentally conscious citizens.

Biodiversity is our environment, and restoring it takes levels of devotion, selflessness and understanding that we as a human race would do well to achieve. Across San Diego, the United States and the world, such activity has begun to emerge, but successfully combating global climate change and the ecological consequences of destruction caused by humankind demands far more. While expensive, the balance between restoration and preservation must shift towards restorative efforts until a healthy relationship between humanity and the environment is reached. Further progress will mean wiping the slate clean and, this time around, using technology to achieve a connection between humans and our fellow passengers on planet Earth that has not been seen for thousands of years.

One solution to saving biodiversity is to expose youth to nature.

Exequiel Ezcurra

We interviewed Dr. Exequiel Ezcurra one afternoon in his office at the San Diego Natural History Museum, where he worked as the director of the Museum's scientific research division, the Biodiversity Research Center of the Californias (BRCC). He also formerly served as provost of the Museum, its highest ranking academic office. Known for his development of Mexico's first environmental impact assessments and most recently for his work on the Sea of Cortez, Ezcurra has published eight books, forty book chapters, and over seventy articles for scientific journals. We were captivated as Ezcurra painted the picture of biodiversity with his eloquent words and his alluring accent.

Student Researcher (SR): What is biodiversity and how does it affect humans and wildlife?

Exequiel Ezcurra: Biodiversity is simply a buzzword we use to describe the heterogeneity of life forms on Earth, the number of different species or the richness of different life forms on Earth. Measuring heterogeneity, measuring the concurrence of different things in one single system, can be a complicated thing but we enclose the whole concept in this buzzword: diversity or biodiversity, which basically means the occurrence of different things living together.

SR: Why is biodiversity significant to any given area?

Ezcurra: In any given spot on Earth, the desert, the tropical rainforests, mangrove swaps, coral reefs, wherever, you will find a lot of different biological species living together, living side-by-side. At the same time, we know that human beings tend to replace systems that are very complex and species-rich with species-poor systems. We get complex systems, we destroy them, and make them simple systems. And you know, for a long time people knew this and nobody thought much of it, but in recent decades we have seen more and more evidence that the destruction of biological diversity is one of the biggest threats to the survival of the biosphere for a number of reasons, for the survival of the living Earth.

First of all, ecological stability. Complex systems have a lot of complex species that interact one with the other in complex ways. That gives a whole system a certain survival ability that tends to disappear or decrease when you replace that system with just one crop. So, biologically simple systems, biologically nondiverse systems, can be very,

very unstable and can lead to immense demographic collapses with dire consequences for the human populations. There is also the knowledge that biological richness is good for us. We obtain a lot of things from biological diversity, even nowadays. Most of the food we eat is derived from wild domesticated crops and most of the animals we eat are from domesticated forms of wild animals.

So, having a reservoir of biological diversity is really good to maintain our natural human management systems. Prospecting in the wild leads to finding new things, especially now that molecular science is so developed. Researchers are finding new enzymes, new genes, or new molecular processes that they can get in the wild by looking at different species that have particular properties that they may deem desirable.

We don't know what role each species plays in the biosphere, and the biosphere is too complex for us to know in a short time exactly what role each species plays. From an authoritative human perspective, even if we assume that we could get rid of unnecessary species and still survive as a world, as a biosphere, we cannot tell which are unnecessary species and which species play critically important roles in the world. So, we have a survival responsibility of keeping as many species as possible, because we actually don't know what role they play in the biosphere.

And then, of course, there is even a more philosophical approach to biodiversity, which is, if you assume an evolutionist's position or an evolutionist's perspective of life on Earth as I have, and you understand that the human species is one species in a context of millions of species in the world, we may be somewhat different just as any species is somewhat different, but we are a part of the biological complexity of the world. Because we have consciousness, we have in some way the moral responsibility of preserving the flow of life on Earth the way we found it. It is an ethical proposition; it is a moral proposition. We have no given right to get rid of other species on Earth merely because we are so powerful and we have access to the energy and resources to do so.

SR: What sort of things influence biodiversity levels?

Ezcurra: Biodiversity is really expressed on at least three different levels, perhaps even more. There is a measure of biodiversity at the genetic level, even within a species. In the wild, this is extremely important because the capacity of a population to respond to changing environments relates to its genetic heterogeneity. The loss of genetic diversity is bad for populations in general; populations need to be different. It is good to have individuals that are different sizes, shapes, colors, that have different metabolisms, etc. That is where natural selection operates, and that is really the engine at the root of survival.

There is another level of biodiversity, which is the population level. Diversity is expressed in the capacity of populations to grow and foster. If you have a species that is incapable of growing for any reason, it will eventually become extinct. Making sure that populations are capable of prospering is very important when it comes to biodiversity conservation.

Then there is another level of biodiversity: the community level, the ecosystem level. In one single ecosystem, the coral reef or the forest, you have a number of species coexisting. All of that is a complex system of things that interact one with the other, and if you eliminate one, the whole system may collapse. So there is a component of biodiversity at the community and ecosystem levels. Finally, of course, there is a component of biodiversity at the landscape level.

So, as you move along the landscape, you will see a lot of different ecosystems; each ecosystem rich in different species. And, of course, the simplification of landscapes through agriculture and development tends also to eliminate species. So, basically, to get healthy ecosystems, we need a lot of variation at the genetic level; we need to maintain genetic variation in populations; we need to maintain healthy populations that are capable of growing; we need to maintain entire ecosystems, or healthy ecosystems, and we need to maintain landscapes—different landscapes. All of those are critical components of a biosphere with viable biodiversity.

SR: What most hinders biodiversity restoration projects?

Ezcurra: All over the world, and I think San Diego Bay is no exception, it is the need for profit and the lack of ability to think more about the general good. What we're basically doing is losing a diffuse service that is good for society at large and converting it to a specific business that is good for a company and the people that use that system. All over the world that is the biggest challenge for conservation. We need to convince society that keeping the larger environmental services provided by nature is important for our collective survival in the long term.

SR: Why is it that people don't believe biodiversity is an issue for concern?

Ezcurra: I'm not sure that people don't believe it is an issue. I find more and more people do—a lot of them. I think it is those people that want to make money that don't consider biodiversity because they have an opportunity for development.

There has been a shift in the perception from the belief that things out there are to be hunted, to the idea that things out there are to be preserved and that preserving them is, in some way, good for us.

Wetlands

For most people, walking through a wetland is a strange and foreign experience. Animal sounds are heard all around, but the sources cannot be determined through the thick grasses. Someone walking on solid ground may take a step and suddenly find himself sinking into water-saturated soil, making much of the area inaccessible. It is no surprise that the alien nature of wetlands has made them difficult for humans to understand and appreciate. Mankind has historically viewed many wetlands, such as the coastal salt marshes of California, as worthless swamps or even hazardous refuges for disease-carrying insects. As a result, hundreds of thousands of acres of wetlands worldwide have been filled and drained for farmland, leaving few of these precious ecosystems in their pristine state. Today, only 3% of the world's wetlands are protected.

The perception of wetlands as useless could not be more false. Wetlands are known for being particularly biodiverse, rivaling even rainforests and coral reefs. Wetlands also provide shelter for a high concentration of endangered and threatened species. Now that conservationists are aware of the ecological significance of wetlands, there exists a large and important environmentalist movement advocating their protection. (Pacific Estuarine Research Laboratory 1990)

A wetland is any area of land saturated with surface or ground water. There are four major wetland types: ponds or lakes, marshes, swamps, and peat bogs. Each one of these has a specific role in our global ecosystem. However, erosion and human interference are slowly making these areas smaller, thereby diminishing many of the wetlands' natural features. Generally, the larger these wetlands are, the larger their environmental role. Carbon sequestration, the

"At Famosa Slough, there are only 37 acres of wetlands, but we have seen over 130 species of birds."
— Jim Peugh

ability to capture and store carbon from the atmosphere, is directly proportionate to the size of the wetland. Although wetlands do not fixate as much carbon as trees, they are still considered to be important to maintaining climate stability. (Spink, Black, and Porter 1991) (Patten, Joergensen, Dumont 1990)

Wetlands also act as an important buffer during times of catastrophic flooding or desiccation. Wetlands assist in flood protection by taking the "peak off floods" by storing water. It is thought that if the wetlands surrounding New Orleans had not been altered, Hurricane Katrina might have had less impact. Additionally, freshwater wetlands may serve as a mechanism for recharging groundwater supplies

Wetlands provide animals with food, refuge, and a site for reproduction. Wetlands are always situated in the basin or lowest point to the surrounding landscape. Many species of animals are instinctively attracted to these low points where food and water are gathered. Since water runs down from higher points in adjacent areas, wetlands become very rich with nutrients. Birds use wetlands as a rest stop on their annual migrations because of the areas' intrinsic protection and plentiful food sources. The plants that border marshes also offer protection for migrating birds looking for a place to rest. As for marine fish, many lay their eggs in coastal wetlands. However, as coastal wetlands diminished, the fishing and shrimp industries were impacted because the animals lost places to lay their eggs. (Cylinder, et al. 1995)

Historically, humans have found much use for wetlands. Such areas were prime locations for hunter-gatherers to settle. During the development of agriculture, early humans found the soil of wetland habitats to be rich with nutrients. As time progressed, humans' relationship with the environment changed from coexistence with nature to exploitation of it. Wetlands became more of a nuisance than an asset. In the late 1700s, a wetland in Virginia was named the Great Dismal Swamp. It was a place "full of nasty little insects" (Greer and Stow 2003), and presented a large problem for travelers. These issues intensified in the early 1800s. As settlers crossed the American frontier, wetlands were perceived as obstacles. Consequently, the government encouraged the elimination of these areas wherever possible. Though our government has made significant progress in their preservation, much work remains to be done to conserve the value

The surviving wetlands of San Diego are rich with wildlife.

of those that remain and restore the nation's severely damaged wetlands. (Greeson, Clark, and Clark 1979) (Spink, Black, and Porter 1991) (Greer and Stow 2003)

In 1960, the U.S. government enacted the "Rivers and Harbors Act" that extended federal jurisdiction to cover all waters within the United States. The original law was presented in 1899, and it is sad that it took 70 years for the law to finally pass through America's lethargic government. By the time Congress addressed this jurisdictional black hole, almost 50% of the nation's total wetlands had been lost. The law enabled officials to prohibit permits for industrial purposes and prosecute "dischargers of pollution." (Cylinder, et al. 1995)

Unfortunately, lawyers were able to find a loophole in the government's definition of a "wetland." To this day, there are still court cases concerning whether or not wetlands fall under the jurisdiction of the government. The most important clause of the Rivers and Harbors Act is that the water has to be "navigable," or used to transport interstate commerce for the federal government to have jurisdiction. Many wetlands such as those around Mission Bay and San Diego Bay are adjacent to such waters and therefore considered protected under the Act. However, many wetlands that fall outside the boundaries of protection are vulnerable to development. These difficulties lead many to believe that the Rivers and Harbors Act is not enough to protect our fragile wetlands. (Cylinder, et al. 1995) (Patten, Joergensen, and Dumont 1990)

It was not until the Clinton administration that wetlands became valuable in the eyes of the U.S. government. Because more than 90% of California's coastal wetlands have been destroyed due to development, the state's wetlands were given a high priority status, which jump-started its wetland conservation policy. There were three goals: first, to ensure no overall net loss and to attempt to gain back wetland acreage in the long run; second, to reduce the complexity of the current wetland administration; third, to encourage restoration through landowner incentives or tax cuts. California uses the definition of wetlands as, "...lands which may be covered periodically or permanently with shallow water and which include saltwater marshes, freshwater marshes, open or closed brackish water marshes, swamps, mudflats, fens, and vernal pools." (California Fish and Game Code §2785 [g]). The laws regulating wetland development apply only to the wetlands that lie within these parameters. Though the government owns some wetlands, the majority are privately owned, which often poses problems. In addition, section 404 of the Clean Water Act helps to protect various wetlands from the discharge of dredged or fill material. Section 1600 of the California Fish and Game Code protects streams to some extent through helping to prevent the diversion or obstruction of the natural flow of a streambed. (Cylinder, et al. 1995) (Spink, Black, and Porter 1991) (Peugh 2007)

Privately owned wetlands are not managed with the best conservation practices in mind. In California's Central Valley, 41% of wetlands are privately owned and thus unprotected. Private duck clubs own many such wetlands. These clubs take conservation seriously and are instrumental in keeping the wetlands healthy. However, the reason that they restore wetlands is often to maintain a thriving hunting ground. Thus, this relationship could bring about problems if these clubs chose to make a profit by allowing the development of their wetlands. (Spink, Black, and Porter 1991) (Patten, Joergensen, Dumont 1990) (Goldstein 1996)

Legislation in California currently states that if an individual receives a permit to develop a place defined as "a wetland," the individual must conserve or restore another wetland to compensate for the loss. Though this policy may at first seem reasonable, it poses many major problems. Human restoration of wetlands can never return a damaged area to its pristine state, and it is always preferable to leave habitats untouched by development. Additionally, acreage is becoming more and more difficult to find and is often great distances away. The Batiquitos Lagoon Restoration Project in San Diego County was mitigation for part of the Long Beach Harbor Project occurring 100 miles to the north. The distance involved is a problem because any animals that were living in the developed wetland had little chance to relocate. In addition, restorative work rarely returns the wetland to its original state. (Spink, Black, and Porter 1991) (Pacific Estuarine Research Laboratory 1990)

California's coastal salt marshes lie along the Pacific Flyway, the annual flight path of many migratory birds. These birds traditionally use such marshes as rest stops, which are vital to the success of their journey. Coastal development and wetland fragmentation is therefore of great concern to the well-being not only of year-round marsh inhabitants, but of migratory visitors as well.

At times wetlands appear to capture victims, perhaps to compensate for the losses the wetlands have experienced.

Though it is clear that wetland restoration work is crucial in California, government renovation projects have not moved with the speed necessary to make serious preservation possible. For example, the Los Peñasquitos Lagoon located in San Diego County is in need of significant maintenance work. A report done by Coastal Environments in 2003 outlined action to restore the lagoon; the two main points were to remove a railroad dike and deepen the channels of the lagoon. These actions would significantly increase the circulation and quality of the wetland. Despite the obvious benefits of the program, it has been four years since the report was issued and the government has yet to fund any maintenance work. Although officials realize the importance of the wetlands, the speed at which government responds to environmental issues unfortunately does not change. (Coastal Environments 2003[1],[2]) (Greer and Stow 2003) (Spink, Black, and Porter 1991)

Famosa Slough Wetland Preserve.

A more successful restoration project in the San Diego area was completed in the Famosa Slough. From 2003 to 2005, the Friends of Famosa Slough led a "wetland surgery" that changed the current of the entire Slough. The Slough did not have any islands, which gave predators, including many introduced domestic species, access to the birds' nesting grounds. In 2003, the project successfully made three acres of new wetland by dredging out fill dirt that had been dumped there decades ago. They did not remove the outer edge of the fill area and this formed an island surrounded by marsh habitat and open water. (Peugh 2007)

Black-necked stilts, avocets and killdeer have nested successfully on the island every year since it was created. Least terns have also been spotted again at the Slough. Not only did the restoration work make the island a sanctuary for birds, it also aided water circulation. The Famosa Slough restoration project was fairly expensive, but was paid for by a grant from the state. Though the government needs encouragement from the public, it can be invaluable in providing financial support for ecological endeavors when enough citizens voice their displeasure and force the government into action. (Peugh 2007)

The solution to saving the wetlands in San Diego Bay is to follow the example of Famosa Slough, which has a volunteer group— "Friends of Famosa Slough." Other wetlands need committees to represent them and if each had a group of volunteers, maintaining and possibly improving wetlands in San Diego would be an easier task. The formation of committees would force the city council to fund more restoration projects, and would also ensure that fewer or no wetlands would be destroyed by future development. This active stewardship is one key component that is necessary to save these remaining critical habitats.

A major wetland restoration project is being planned by the U.S. Fish and Wildlife Service in South San Diego Bay. It should result in hundreds of acres of new, highly productive wetlands. But currently, there are no funds to implement it. (Peugh 2007)

The problem of saving our wetlands is a convoluted issue

because there are often many forces working against their preservation. While development can be seen as the most devastating factor, many lesser-known issues have become increasingly detrimental. Since wetlands are often the lowest-lying regions of an area, even upstream human activity causes erosion and pollution of the beautiful marshes. Over time, as people water their lawns and wash their cars, wetlands are slowly being filled by the runoff from these various activities. After erosion has played its devastating role in the destruction of a wetland, it may become necessary to attempt to restore it to its original state. This option is expensive, but may be necessary to save the entire ecosystem. In order to combat erosion, a fee could be paid by developers of a watershed that would be used towards wetland restoration. Because of erosion problems, all development projects are required to use containment methods to prevent silt runoff coming from construction sites. It is time to forget about the "insignificant" financial conflicts and worry about the future and well-being of our environment. (Kusler 1990) (Pacific Estuarine Research Laboratory 1990)

In terms of economics, wetlands in the United States are estimated to account for as much as several billion dollars of profit. As these precious areas diminish, these benefits decrease. That leaves us with only one current option: costly and time-consuming reconstruction of wetlands. However, the majority of wetlands are gone and we cannot afford to lose any more. (Cylinder, et al. 1995)

In the past, the U.S. government actually encouraged the destruction of wetlands or what the legislation called "swamps." The U.S. Department of Agriculture was the governmental culprit behind their destruction because they wanted to promote agriculture. Wetlands, being at the bottom of a drainage basin, had several characteristics that defined them as prime agricultural land: flat topography, rich topsoil and excellent sources of water. (Greeson, Clark, and Clark 1979)

Unfortunately, some of these same criteria make for prime real estate and consequently, many large U.S. cities (or large portions of them) are located on filled-in swamps: Washington D.C., San Diego, New Orleans, San Francisco and Boston are a few. Philip Greeson describes this mentality with brilliant simplicity, "We have devoured our wetlands the way we eat potato chips—never stopping at just one." (Greeson, Clark, and Clark 1979)

Human activities have slowly been making wetlands smaller; large bulldozers replace serene landscapes with large condominiums. In 20 years we will see one of two things: more large buildings or peaceful birds nesting in their wetland habitats. Each individual needs to get active and join the fight for their local wetlands in order to preserve one of our most vital local ecosystems. (Pacific Estuarine Research Laboratory 1990) (Spink, Black, and Porter 1991) (Greeson, Clark, and Clark 1979)

Jim Peugh

When one thinks of conservation work in San Diego Bay the name "Jim Peugh" immediately comes to mind. Peugh has made amazing contributions, working at Famosa Slough and Mission Bay wetlands. In his efforts to improve wildlife habitats in San Diego, he consulted the city to look past the financial tribulations and restore wetlands regardless of cost. Peugh is currently the Chairman of the City of San Diego's Wetland Advisory Board, Chairman of the Friends of Famosa Slough, and a Conservation Chair for the San Diego Audubon Society.

We met Peugh on a warm, brightly lit San Diego day at Famosa Slough. He seemed to radiate knowledge about the wetland where we were standing. His soft-spoken voice gave a sense of calm that made us believe he was a part of the nature around us. As Peugh continued to speak, birds slowly swam by and merrily bathed in the clean water. Their playful sounds carried around the wetland sanctuary. Just a short walk away, cars continued to whiz by us, but while in the wetland it felt as if we were sheltered from the stressful life of our hustle-bustle society.

Student Researcher (SR): What exactly are wetlands?

Jim Peugh: Wetlands are simply areas that are dominated by water, but are defined by different agencies in different ways. With the U.S. Army Corps of Engineers the definition is very stringent. It has to do with the kind of soil, the amount of water and the kind of plants. The definition of some agencies, like the Coastal Commission, has to do with plants— if it has wetland plants on it, then it's a wetland. There are wetlands out there that don't get protected because they don't fit the exact definition of the agency that is responsible for them.

SR: How are wetlands important to plant, animal, and maybe even human life?

Peugh: For human life, they're really a nice place to be. They're tranquil, they're pretty, and you can learn a lot about nature from them. For wildlife, they're just as important because many species depend on them. Here at Famosa Slough, there are only 37 acres of wetlands, but we have seen over 130 species of birds here. Wetlands also provide nurseries for various fish. A lot of the fish people make money off ocean catches wouldn't exist unless wetlands were here.

SR: What is being impacted the most by the loss of wetlands?

Signs are one method used to promote the preservation of wetland habitat.

Peugh: Fish in general are severely impacted by the loss of wetlands. Water quality also is impacted by the loss of wetlands. Water that comes off our watershed and then through the Slough is a lot cleaner than water from Ocean Beach that gets dumped straight in the flood-control channel and goes out to the ocean.

SR: How does a wetland "clean" the water specifically?

Peugh: It can be a physical filter. The dirt that comes in it gets stopped and drops out; trash gets caught. Some petroleum and other pollutants are broken down chemically by the bacteria that's in the soil. The excess nutrients in the runoff are used by the wetland plants.

SR: If wetlands are so valuable, why are our wetlands being destroyed?

Peugh: The rate at which wetlands are being destroyed is much lower than it was 40 years ago. But wetlands are still lost for construction of new roads or pipelines or to expand old ones. If developers have properties with wetlands, they often try to see how much the agencies will allow them to build on. If the agencies give in, the developer will build to increase their profits. Also, the U.S. government has made uncertain the Clean Water Act's protection of isolated wetlands and intermittent streams, which will result in more construction on wetlands. The agencies that protect wetlands are typically under funded, so they do not have enough staff to uniformly implement our protections for wetlands.

SR: Are there any areas of San Diego Bay that represent what wetlands were like before European settlement?

Peugh: The best place in San Diego Bay to look for traditional wetlands is along the Silver Strand. There is something called the Biological Study Area, which is a Navy property that is being maintained by the county as a wetland. There's also the Sweetwater Marsh, at the mouth of the Sweetwater River, and the Tijuana Estuary.

SR: Are there laws and regulations that have been put into effect to preserve wetlands, and have they been successful?

Peugh: There are a lot of laws out there to protect wetlands. However, some of them are successful, and some of them aren't. It's on an event-by-event basis. The most effective protection for wetlands is in the coastal zone. The Coastal Commission is tougher than other agencies when it comes to protecting remaining wetlands. The least successful are projects that are related to the Corps of Engineers. They tend to slip by a lot of the wetland laws because they have been weakened by politics. But, the thing we have the most problems with is what's called isolated wetlands and higher order streams.

SR: How can we take an active role in the conservation effort?

Peugh: You can do everything. You can do as much as I can. If a new face shows up at the city council or the regional board meetings and speaks up, this has a lot of impact. You not only need to think about being an activist and educating politicians and trying to raise money for these things, you also need to be a responsible citizen. When you see something you don't like, call a politician and tell them. When you read an editorial in the newspaper that's really silly, call the newspaper and tell them. They need to know that they can't get away with the kind of silliness that we're being fed right now. You can contribute a lot by doing physical work with some organization, like removing weeds, planting native plants, picking up trash and preventing erosion.

NATURE'S UNEXPECTED SPLENDOR

A spread of muck sloshing under my feet.
I wish I could be back near civilization, civilized life.
I think, "What could live in this desolate land?"

A rustle behind me and my heart skips a beat.
I turn to see long elegant legs
Of an elegant bird dipping under the surface.
Why such beauty in a brown, groggy place?

She turns to me with her beady eyes,
Dips her head low, and to my surprise
She pulls up a blade of grass and is eating it whole.
The bird, with her elegance, must live in this wet land.

Far in the distance, I see its grand splendor:
An ecosystem of life hidden 'neath the first glance.
For hundreds of feet till I see the city skyline.
Through muck and water,
I see the creatures *surviving*.
Beauty is here.
Life is here.

— Megan Morikawa

Shifting Baselines

Remember the childhood game of "Telephone"? One child invents a phrase and whispers it to the child next to him or her. The process continues until it reaches the end, where the final player recites what he or she heard from the preceding player. In most cases, the original phrase is significantly distorted due to accidental misinterpretation. But how does that relate to conservation? It is actually the same general concept as a major ecological problem called "shifting baselines," which many believe is one of the most important problems in conservation biology. Similar to a phrase in "Telephone," but on a much larger scale, these global changes have a similar effect. Whether the cause of change is natural or sociological, the change is there. To understand the concept of "shifting baselines," one must understand that a baseline is, in essence, a point of reference to which other things can be compared. The "shift" describes the baseline gradually changing over time, usually through generations and usually unnoticed. Because of this gradual change, the initial state, the "pristine baseline" is forgotten and lost.

Dr. Jeremy Jackson of Scripps Institution of Oceanography has based much of his recent career on publicizing the concept of "shifting baselines." He focuses on the generational shift by using an example of one generation versus the generations before it. "Shifting baselines is the idea that what we think of as natural in the world is a function of our experiences as kids. Because of this, we assume that nature is the way it was when we were children; and as we grow older and see it change, we get upset. We see these changes, but it never occurs to us that the same thing has happened in all the generations before us. So, none of us really has an idea of what nature is and how much it has changed ...." (Jackson 2007) In this way, Jackson demonstrates how the cycle is unconscious and continuous, changing

"Shifting baselines is the biggest problem of the environment as far as I'm concerned.... If people don't know what they've lost, then why should they care?"
— Jeremy Jackson

our idea of what is "pristine" with every generation.

The history of the term "shifting baselines" is recent, developed in 1995 by marine biologist Daniel Pauly. The concept itself, however, is timeless and can be applied for any "shift" throughout history. The term is usually used for unconscious shifts that occur over long periods of time. There are many well-documented observations of past populations of oceanic species during the age of exploration and earlier that provide evidence of baselines that have shifted. For example, eighteenth century explorer George Shelvocke wrote in *A Voyage Round the World by Way of the Great South Sea* of whales that were so abundant, he was actually worried about ramming them as he sailed. After centuries of excessive whaling, the population levels that Shelvocke described have not been seen for centuries and even seem unrealistic by our standards. Therefore, while it is easy to recognize a drop in whale populations, it is difficult to precisely quantify the damage because it has occurred over numerous generations. (Pauly 1995) (Shelvocke 1726) (Jackson 2007)

The original example that inspired the term shifting baselines was an examination of fisheries by Daniel Pauly. He discussed how the "standard" population of fish was re-determined by the beginning of each new generation. Pauly described a large, gradual decline of fish population that went nearly unnoticed because of the generational gap. Thus, over the years the perceived pristine state of fish populations was greatly distorted. (Pauly 1995)

Pauly's article on overfishing was a catalyst that inspired the start of the movement towards fixing the shifting baselines syndrome.

The term was soon recognized on the same level as large-scale environmental issues such as global warming. It was realized that this was a problem that could only be addressed through widespread awareness and the combined efforts of several different interest groups. In this case, the cooperation of environmentalists, scientists, industries and the public was necessary. The process began when several concerned colleagues founded shiftingbaselines.org. Marine ecologists Randy Olson, Jeremy Jackson, Paul Dayton and Steve Miller collaborated with two movie producers, Gale Anne Hurd of Valhalla Motion Pictures and Jason Ensler of NBC Movies, to create a website addressing one of the most important components of shifting baselines: public awareness. (Olson 2007) ("The Team" 2007)

The kelp forest past and present lenticular display produced by Shifting Baselines clearly demonstrates the loss of species abundance and diversity.

Before establishing shiftingbaselines.org, researchers analyzed the results of various phrases in online search engines. "Overfishing" came in at 189,000, "marine protected areas" at 782,000, "Jeremy Jackson" at 670,000, "baby seal" at 1,050,000, and "shifting baselines" at 16,200. From this it could be determined that internet users were 60 times more interested in the status of baby seals than that of the oceans in which the seals lived. Thus, the first step towards the future of shifting baselines was to spread awareness of the seriousness of the problem on a public, national and international level. Shiftingbaselines.org began their focus by addressing publicity through amateur and professional public service announcements and comedic videos. By shedding light on the issue, the founders of the organization hoped to at least familiarize the public with the term. When the search-engine test was performed again on August 13, 2007, shifting baselines had 352,000 hits. This revealed that the movement was gaining recognition and might be educating many more people about the status of the oceans. ("A Worldwide Trend: MPAs" 2007)

The next phase in the project deals with restoring the environment to an original or healthy baseline. This is the largest and most difficult step of the process. Even before restoration can occur, an "original" or sustainable baseline must be determined for any given species so that case-by-case goals can be set. The process could begin by examining the previous baselines and environmental status of a certain ecosystem, yet this would be limited by the amount of research and history that has been recorded in that area. A secondary step could focus on the gradual reintroduction of lost species in order to revert to the predetermined, "optimal" baseline. One way to think of this step would be an attempt to remove a species from the Endangered Species list. Later in this book, the many factors involved in this process are discussed. Regardless, a comprehensive understanding of the ecosystem and all factors affecting it are needed to even begin to revert to a sustainable baseline.

In San Diego, distinct examples of shifting baselines are all around us. Take a moment to imagine San Diego Bay in its pristine state. This might be before the first settlers, even before the Native Americans came to the area. Think of coastal wetlands; the closest dry land was where Old Town is located now. The rest was untamed marsh. The San Diego River would flow one way one year, then another the next. It was wilderness with more wildlife than could ever be imagined today. If we compare this past with the condition of the modern bay, the images would be altogether unrecognizable. The baseline has shifted many times over the centuries since Europeans arrived. (McKeever 1985)

As history progressed, the process continued, each civilization slowly altering the natural ecosystems around them with new

A comparison of the San Diego waterfront c. 1890 and 2007 illustrates an example of a local shifted baseline. 1890 photo: San Diego Historical Society.

innovations and sociological changes beginning with Cabrillo's arrival in 1542; then the Spanish settlers; the Mexican Revolution; the introduction of stock animals; the 1847 Mexican-American War; which led to the deepening of San Diego Bay to allow the passage of more ships; the seaside industry dumping runoff freely into the bay; the arrival of the Navy in the twentieth century; the commercialization of the fishing industry, and much more. Over this period, anything and everything was done to facilitate the growth of business in San Diego. In the name of progress, an entire irreplaceable ecosystem was lost forever. (For more history, see "Ship Traffic" and "Land Use.") (McKeever 1985)

Today, we see San Diego as a thriving and developed city, and we know it was not always like this. But if we were to revert to an original baseline, what would we choose? Does anyone remember the marshlands stretching from the coast through all of San Diego Bay? Can anyone imagine the giant kelp forest offshore stretching towards the horizon for miles? Can anyone still picture the multitudes and diversity of fish swimming in and birds flying over vast natural wetlands? How do we even begin to remember? How can we know which baseline is "correct," as well as sustainable?

Once we know—and we must know soon—it is necessary to make changes on various levels in order to preserve what is left in San Diego. A change of mindset is the most important step. One way to initiate change is to set aside habitat and allow the world's natural healing to take place. We have taken steps to preserve the existing natural habitats in and around the bay, and even some towards restoring it to something like its former self. This is shown in the recovery of habitats such as those at the Chula Vista Nature Center that has been converted from a thriving salt marsh, to an industrial manufacturing facility, to an area used for commercial agriculture and finally, back to a thriving salt marsh.

Messages similar to those of Dr. Jane Goodall and those who manage shiftingbaselines.org show that San Diego cannot and should not lose all hope. What would the world look like if no efforts are made to stop the current trends of shifting baselines? Scientists like Dr. Jeremy Jackson believe that, in 50 years, the oceans may become dismal wastelands if current trends continue. How do we prevent that from happening? Solutions such as Marine Protected Areas, Stewards of the Bay, sustainable fisheries, and various levels of recovery (see "Solutions" section of book) could be the answer. But first, we must become aware of the problem and understand what "pristine" truly means. Once the problem is recognized, then we will finally be able to move towards a solution. (Jackson 2007)

Jeremy Jackson

Dr. Jeremy Jackson of Scripps Institution of Oceanography was interviewed on a sunny day next to the beautiful vista of La Jolla Shores. The true beauty and state of the ocean was a hot topic in our interview, which centered on the concept of "shifting baselines." Dr. Jackson is the William E. and Mary B. Ritter Professor of Oceanography and Director of the Center for Marine Biodiversity and Conservation at Scripps, as well as author of more than 100 scientific publications and five books. As an acclaimed marine biologist/ecologist, he has received many awards commending his research. Despite the recognition, Jackson's main passion lies in ecological and environmental awareness, a concept that he explained in great detail in our interview. As cofounder of shiftingbaselines.org, a website targeted at young audiences, Jackson truly shows that the idea of conservation and ecological awareness must be inspired and portrayed among our youth.

Student Researcher (SR): What is shifting baselines?

Jeremy Jackson: Shifting baselines is the idea that what we think is natural in the world is a function of our experiences as kids. We just automatically assume that nature is the way it was when we were children. As we grow older and see all sorts of things change, perhaps we get upset about it. We see that change, but it never occurs to us that the same thing happened the generation before ours and the generation before that, so none of us really have an idea of what nature is.

SR: What are key signs that show a shifted baseline?

Jackson: Probably the most obvious is when you go out in the ocean and try to see a fish that's bigger than a foot. Every once in a while we get beautiful dolphins coming in; every once in a while we see whales off shore, but out here in the kelp forest there used to be these big, enormous, giant sea bass that were eight feet (2.4 meters) long and weighed many hundreds of pounds. Nowadays, you see one fish that big in a year of diving and you say, "Oh my goodness, that was a really amazing thing we saw." But Paul Dayton, a scientist at Scripps, has a whole collection of photographs from when he was here only 30, 40 years ago of spear fishermen coming in with tens and hundreds of these fishes. Not having big fish is a really obvious sign that things have changed.

SR: How can you project how a certain environment appeared when it was in a pristine state?

Jackson: It's really a hard thing to do. I mean, first of all, the world is always changing naturally anyway. There were ice ages and now we're not in an ice age and that doesn't have anything to do with people. In trying to understand what the world would be like without us, we have to try to understand all those things and then ask the question: "Well, how is it different because of us compared to all those natural things?"

SR: Are there terms in ecology or conservation biology for the idea of shifting baselines that were used in 1990?

Jackson: Shifting baselines is a saying. I mean people talk about degradation and change, but there's something about that expression that is catchy. People say, "Well what does that mean?" It's really important to try to convey ideas about worlds that are different and strange and hard to understand in simple ways that catch people's attention. Shifting baselines catches people's attention. When you say it's just the idea that the world has changed so much that we don't even know what it was like, people say, "Oh, that's very interesting." Then they ask questions, and it draws them in.

SR: How can individuals help?

Jackson: Social movement happens because people really care. Just the way we try to keep shifting baselines going, they never forget that they really care. They think about how they vote, and they think about how they interact with the people they know, always remembering this issue that they really care about. So, the most important thing that anybody can do, and especially young people, is to do something that sounds really corny,

In the early days of San Diego spearfishing, large fish like this white seabass were plentiful.

and that is to be good citizens. To actually think about the world you live in and ask whether or not you like what you see going on in the world. If you don't like it, what can you do to change it? I can assure you that if millions of people wrote to their government officials with opinions on aquaculture, or overfishing, or Marine Protected Areas, and these officials got 10,000 letters a week, then all of a sudden they would really care. But, if nobody ever writes them, and nobody ever pushes them, then all the other things that people push about are what will get attention. Things change because people really, really care. So what you can do is really, really care, and have the courage of your convictions, and live a life of your convictions.

SR: You talk a lot about shifting baselines specifically in the ocean. How do you feel that shifting baselines relate beyond the ocean?

Jackson: Shifting baselines are the biggest problem for the environment as far as I'm concerned, because if people don't know what they've lost then why should they care? I grew up for six years of my life in New York City and, in a strange way, I believe people in New York have a better idea of what's going on in the environment than people who live in a place like [San Diego]. They don't notice all of the things that are happening, but if you live in Manhattan and you plan your trip to get out of Manhattan to go to some beautiful place in the country, you actually really think about it. Every day, the degradation happens a little bit, little by little, and we don't notice.

SR: You were talking about the 1930s and going back maybe eighty years or so; how far do you think we should go back looking at issues such as shifting baselines?

Jackson: It depends on where you are. There are big stone tablets in the Louvre in Paris that are about two meters wide. They show two boats in the Mediterranean pulling cedar trees—the famous cedars of Lebanon—up to Turkey so they could be dragged to Persia to build the palace of Darius the Great. In these tablets, all around the boats there are these amazing gigantic sea turtles and all this stuff you could never see today. Why did they put that stuff in there? Because that's what it used to be like. That was 3,000 years ago; the Mediterranean was the cradle of Western civilization. People had been screwing up the Mediterranean for about 5,000 years., In contrast, people had comparatively little impact in the Caribbean until about 1,000 years ago. In Australia, the aborigines didn't use boats to fish, so that large impacts [in fishing] did not occur until 200 years ago when Europeans arrived and settled.

SR: How have shifting baselines affected San Diego Bay?

Jackson: Sometimes I go up to the [Birch] aquarium, and I see people looking at the lenticular displays of shifting baselines and just watch them or that ghost forest tank and the kelp, and I think there's an increasing awareness here. I think San Diego Bay itself, as far as I can tell, is a pretty terrible mess. Yet, it's got green turtles in it that have migrated here and hang out in that warm water from the power plant. I don't think we would really like to know what all the toxins are that have been dumped into that bay from the naval facility. If it's anything like any other naval facility, you can be sure there's some pretty horrible stuff in that water. It obviously bears very little, if any, relationship to what it was like before. It's a totally different world.

SR: So what do you think is natural in San Diego Bay?

Jackson: Oh, I'm not sure anything is still natural in San Diego Bay. I don't think there's anything natural on this coast, except maybe the beach and the cliffs down there. Look at the argument around the Children's Pool, and the seals, and kids. That's a totally artificial environment that has been adopted by the seals and hopefully, people will leave the seals alone. It's just really nice they found a home there, but it will never be the way it was. It will never be natural. A guy named Bill McKevin wrote a book some years ago called, *The End of Nature*, in which he pointed out that no place on Earth has escaped the rise of CO_2, so that no place is pristine. But there are some places that are more natural than others and that's probably a good thing.

The birds of the bay that have resided here for millenia now must adapt to a habitat that is starkly different from that of their ancestors.

BREATHE ADIEU

I watch the tides upon the sand;
They shape the gently sloping shore.
The meeting of the sea and land:
How did this place appear before?

The waters of the bay have seen
The impact of a drastic change.
What could the wordless whispers mean?
Familiar still, and yet so strange.

So rarely now is nature found
Untouched by damage of mankind.
While poisons dwell beneath the ground,
We leave Earth's treasures far behind.

With sadness now I face the truth:
Man brings with him an endless pain.
Vivacity of nature's youth
Will never here be seen again.

— Sean Curtice

DNA Barcoding

The identification of species is vital to understanding life on Earth. Intricate differences within groups of plants, animals, fungi, protists and bacteria define their roles in the environment, and knowledge of the scope of life helps scientists evaluate biodiversity and ensure that ecosystems are properly sustained.

For centuries, species were established and recognized entirely by their morphology (physical appearance). In the mid-eighteenth century, the legendary Carolus Linnaeus (1707–1778) revolutionized the discipline of biology through the development of a system for naming, ranking and classifying organisms. Linnaean taxonomy, as his creation was named, organized life through a hierarchical system that used Latinized names to organize species into increasingly small groups. Each species was given a specific name through the system of binomial nomenclature that designates the genus and the species. For example, humans are known as *Homo sapiens* (genus *Homo*, species *sapiens*). Linnaean taxonomy is still the scientifically accepted system for biological classification today, nearly 250 years after its inception.

Though morphological taxonomy is often effective for classifying an entire specimen, the same is not always true when attempting to identify the origin of a small sample. Commonly feathers, fecal material, residual tissue or meat samples cannot be identified through a classical taxonomic approach. Thus, novel methods of identification using molecular biology have been developed. Deoxyribonucleic acid (DNA) is, for the most part, unique to every species and even every individual, and is present in all living cells of every organism. The unique nature of DNA is analogous to the human fingerprint and, much like that of the human, DNA can be used to identify species by understanding its structure and properties. The "building blocks" of DNA come only in four types or bases: adenine, thymine, guanine and cytosine. The arrangement of these bases in unique sequences makes up genes that determine physical traits. The unique sequences can be used to identify a species, as well as differentiate individuals within a species. Just one strand of DNA can outline the identity, the "DNA barcode," of any living species.

Developments in DNA science have also drastically changed phylogeny, the science of how organisms are evolutionarily related. While knowledge in this area was previously based on morphology, fossil record, embryology, biochemistry and biogeography, DNA sequences have made direct genetic comparisons of species possible. This has allowed mankind a much deeper understanding of the history of life on Earth.

At High Tech High, students in Dr. Jay Vavra's biotechnology class study genetics in the form of conservation forensics. Vavra's quest to bring the science of DNA barcoding to his classroom began through collaboration with Dr. Oliver Ryder from the Conservation and Research for Endangered Species (CRES) Center of the Zoological Society of San Diego. Much of the work in the conservation forensics class focused on combating the African bushmeat crisis, the illegal meat trade from protected species that is often disguised as meat from a lawful origin. High Tech High students practiced and developed methods for extracting, isolating, amplifying and sequencing DNA from dried jerky samples, which simulated the difficult-to-identify samples that might be found in an African marketplace. The hope was to use the resulting DNA sequence to work in conjunction with partners such as CRES, Invitrogen, the Smithsonian Institute, and the Bushmeat Crisis Taskforce to develop a strategy for combating the bushmeat trade. In this first-ever High Tech High DNA project, students were able to effectively verify the origins of ostrich, turkey and beef jerky. With the project's successful methodology and collaboration, the students involved took the torch of DNA science and utilized this method of species identification to address conservation issues in San Diego Bay, ranging from plankton identification to studies of invasive species. (Vavra and Ryder, 2006)

An organism's entire genetic sequence or genome is incredibly long: that of humans is about three billion base pairs. The specific sequence of DNA used in DNA barcoding is found in the mitochondria of a cell. The mitochondrial genome is much smaller than the nuclear genome at about 15,000 base pairs. Though the majority of DNA in an animal cell is found in the nucleus, mitochondrial DNA is most useful for evolutionary studies and applied species identification because it is inherited maternally. This is unlike the nuclear DNA, which is frequently scrambled in sexually reproducing creatures since the maternal half of the DNA is recombined with the paternal half of the DNA in each generation. The mitochondrial gene primarily used for identifying animals is called Cytochrome c oxidase, subunit I (COI) and is approximately 710 base pairs long. Interestingly, this gene is critical to the conversion of energy that takes place within the mitochondria, and without its normal function an animal would not be able to live. It is not the function of COI that is important in the DNA barcoding process, but rather the sequence of a portion of the gene that can allow one to identify a given species, because the sequence is different between species.

With regards to conservation biology, DNA barcoding is amenable to the identification of species through trace samples. This method of noninvasive sampling contributes to conservation efforts on many levels. Trace samples allow scientists to identify species using feathers, feces, hair follicles, egg shells, ivory and other matter. Because trace sampling can be obtained without harm to the specimen and does not require that a living organism is present, it can be used to help survey and control transportation and trafficking in the illegal meat market.

As previously mentioned, the bushmeat trade is a key concern for the conservation of endangered species throughout Africa. Illegally hunted animals often end up in markets, mislabeled as legal meats (see www.Africanbushmeat.org). Unfortunately, processed meat from illegally hunted animals is virtually indistinguishable from the meat of legal game. Similar problems exist in other parts of the world as well. In Japan, the trade of whale meat known as *kujira* faces similar difficulties with poaching and trafficking. The International Whaling Commission (IWC) places strict quotas on the number, species and location of whales acceptable for trade. Biologists Stephen Palumbi and Scott Baker decided to investigate whale meat to determine how much actually came from legal minke whales. "If you can't tell the difference between a minke whale and, say, a fin whale when it's wrapped in cellophane, how can you police that? That's where the genetic technique is really powerful," says Palumbi. The applications of DNA for species identification remain highly valuable in an array of situations. (Palumbi and Baker 1994) (Lorenz et al. 2005)

The science proved to be so strongly applicable to conservation in so many ways that we decided to develop our own in-house method of DNA extraction. One of the strongest arguments against DNA barcoding as a solution for trafficking illegal meats or a primary means of identifying species is the amount of equipment, technology, knowledge and funding required to process the delicate material. At High Tech High, we strove to establish a simplified approach that could be performed in our school's biotechnology laboratory. Since the project began, High Tech High's DNA barcoding process, with the help of CRES, has optimized and refined the steps involved in the isolation, extraction and identification of species through a specific DNA sequence.

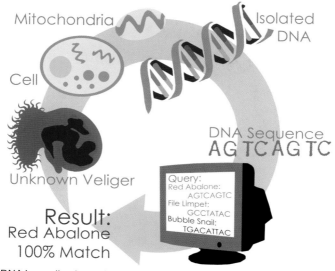

DNA barcoding is useful for determining the identity of species that are not easily recognized by their morphology (e.g., veliger—the final larva stage of abalone) or from which only trace tissue samples are available.

The first step in the DNA barcoding process is to add a detergent solution to lyse the cells by breaking down the plasma membrane of the cell wall and exposing the organelles of the cell. Once the cell's inner components are exposed through lysis, the rest of the components are also destroyed and the DNA is isolated through a proteinase extraction and ethanol precipitation. The proteinase denatures the unwanted protein and the ethanol assists in precipitating or isolating the DNA. After checking the purity of the process with a spectrophotometer, a device that uses light of varying wavelengths to measure the concentrations of protein and DNA, if an adequate level of purity is achieved then the DNA sequence of interest can be amplified. Once the pure DNA is obtained, the 710 base pair region of the COI gene can be amplified. This amplification is achieved using a method called polymerase chain reaction (PCR). In this process, an enzyme called DNA polymerase is added to make new DNA from the existing template in the sample of choice. This is achieved with the addition of deoxynucleoside triposphates (dNTPs) (A, T, G and C bases), buffering salts that protect the DNA, and primer DNA. COI is useful for DNA barcoding because it has a variable region inside conserved ends. The different sequences of COI could be thought of as a vast collection of fairy tales, each of which begins with the phrase, "Once upon a time ..." and terminates with, "... happily ever after," but with a different story in the middle for every species. Specifically, primers are short sequences of DNA that identify where the DNA polymerase enzyme should attach to allow DNA replication, in this case the COI gene.

Various primers attach differently to different species, so a multitude of primers were used throughout the course of the barcoding class. Since related species have similar DNA sequences and structures, the same primer may be used for closely related species. For example, a primer tailored to mammals may react to chimpanzee and field mouse DNA but might not amplify lizard or fish. Primers are short segments of single-stranded nucleotides of about 20 base pairs in length. Vertebrate primers from a related primate study (Lorenz, et al. 2005) were used in the bushmeat simulation with the jerky samples. Research later led the study to a scientific paper (Folmer, et al. 1994) that identified more universal primers that were necessary for the range of taxa involved in this current project. This allowed the identification of arthropods, mollusks, cnidarians and chordates.

After PCR, the samples must undergo another process to be viewed by the naked eye to verify successful amplification of the sequence of interest. This process is called gel electrophoresis, and is used to separate DNA fragments with electricity in a thin slab of gel matrix made of agar. Since DNA is naturally electronegative, an electric current will pull the negatively charged DNA through the gel. Because only the short gene (in this case COI at approximately 710 base pairs) is abundant, these smaller fragments will run through the gel faster than the long strands of genomic DNA. The DNA in the gel is stained with ethidium bromide that fluoresces when exposed to ultraviolet light. This step allows the visualization of the PCR product(s) and verification of the size of resulting DNA to determine if the PCR successfully amplified the COI gene.

Once gel electrophoresis verifies the DNA amplification, the next step is to sequence the DNA. This was done using chain-terminating sequencing reactions, which labels the DNA polymer with fluorescent bases of four different colors to visualize the DNA sequence after the DNA is separated with another round of electrophoresis (capillary electrophoresis). The results from sequencing produce an electropherogram that reveals the sequence of As, Ts, Gs and Cs that compose DNA. To determine the species of origin, the electropherogram is reduced to a sequence and aligned with all previously published sequencing in GenBank, using an online program called BLAST (Basic Local Alignment and Search Tool).

In our laboratory, PCR takes more than three hours for as few as 24 samples, but advanced nanotechnology can perform millions of PCR reactions in under a minute. Complex chemically bound paper allows one-step transformation of raw, field collected DNA to isolated DNA. Students still have to go through many steps individually. Yet the comparison is similar to the differences between an artisan and a factory worker. While our technique may achieve DNA barcoding in a more modest manner, it requires a great deal of knowledge and skill from the students involved. Because of this, it not only leaves the potential of replicating a simplified approach for the bushmeat trade, but allows our classes to continue an important study on genetics, phylogeny, and the impact of species on the environment.

In the second semester of our year-long study, we attempted to apply our methods directly to conservation in San Diego Bay. Three series of classes each focused on a new approach, unveiling discoveries to aid the process of DNA barcoding for each class. The first of the three classes focused on plankton identification. Many creatures of the ocean begin their process of life in a planktonic (free floating) stage as embryos or larvae, and then later develop into their adult forms. In an embryonic stage, visual or anatomical identification of creatures proves to be extremely difficult (even for humans as human embryos look similar to that of pig and salamander embryos). The abalone is a species of concern that is a perfect example of this problem. In its larval form in the plankton, the abalone veliger larvae look very

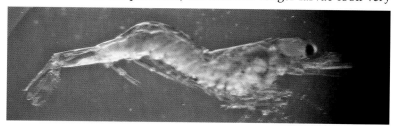

Students engineered a remote-controlled boat to operate a plankton net for collecting specimens to be barcoded. The zooplankton shown above was successfully identified by the students as a local *Euphausid*.

similar to a variety of other gastropods that include: limpets, snails, barnacles, and many other widely abundant mollusca. We hoped we could use DNA identification to help with this problem.

The second of the three classes set about to broaden the species range of samples and attempt to amplify DNA from more delicate trace samples. The universal primers (HCO and LCO) from the aforementioned Folmer, et. al 1994 study allowed our class to create successful PCR reactions for a diverse set of animals. From reptiles to human hair follicles, a variety of mediums were tested, yielding positive results in many samples. With efficient methods and versatile primers, we now had the technique to increase the amount of samples and focus their relevance on conservation in San Diego Bay.

The third class began by taking a trip down to the San Diego Bay Boat Channel to collect samples for barcoding. Samples of the suspected Asian mussel (invasive species) were taken to verify if its visual identification matched its DNA barcode; samples of the local bay mussel were taken and sequenced to verify the specific strain of species that live along the bay; samples of unidentified floating sacs (egg masses) were sampled to support identification of species in this region for biodiversity; and samples were taken with noninvasive swabbing techniques to support the idea of trace sampling. Each sample had relevance and had the primers and methodology to help sequence the DNA. All that was needed were positive results to see whether the true barcode matched the hypothesis. Similar to the other classes working with DNA barcoding, we processed the samples in the HTH Biotech lab to the point of PCR and gel electrophoresis. We were able to achieve successful PCR reactions on a variety of specimens that will be described in this book. We turned to our partners at CRES to take the final step in determining the DNA sequence.

As the samples were sent off to be analyzed by our collaborators at CRES and their DNA sequencing machines, we waited to see whether our hypotheses were correct. There were some true unknowns of species that we believed might never have been sequenced before. In addition, through species identification, we could verify the specific genus and species of native and introduced creatures of the bay. If our hypotheses were correct, we could potentially make true contributions to conservation biology—an exciting feat for mere high school students. When the results were received, we entered the sequences into an online database for genetic sequences on the National Center for Biotechnology Information website to confirm the species.

Results yielded exciting news as test after test projected much wanted information for the majority of samples analyzed. The Asian mussel gave positive results, verifying the presence of an introduced species and creating a DNA barcode to identify the species. The bay mussel was further verified to its genus and species, information which was initially unknown between three close species of bay mussel. DNA extracted from the shells of local mussels further told us that trace samples of recently deceased mollusca were sufficient to identify the species through DNA barcodes. Thresher shark meat from a local fish market verified that the label did in fact match the meat described on the package, suggesting that this market was labeling their products correctly from catch to consumer. And finally, the unknown gelatinous sac showed strong indications of belonging to an annelid or segmented insect local to the bay's sandy shores. Our results were tremendous.

Fecal samples of raptors, including osprey, burrowing owl, and peregrine falcon were unsuccessful, most likely because of excess contamination from dietary items. It is hoped that future work will improve the ability to barcode from fecal matter since animal surveys could be conducted simply by analyzing these noninvasive samples.

While our results were a great achievement for our biotechnology class, our contribution to conservation remains small in the massive efforts needed to quell the detrimental dangers facing species and ecosystems around the world. The future of our project to incorporate genetics remains hopeful; we currently have plans to take our studies to Africa and directly combat the bushmeat trade. For now we can only stress the idea that species identification through genetics holds promising hope for the future of conservation. In the words of E. O. Wilson, "… each species is a masterpiece, a creation assembled with extreme care and genius." It is our hope that our endeavors into the world of genetics help unfold these delicate masterpieces and preserve their vital role in the environment.

ORGANISMS

"If you save the living environment, the creation, the fauna and flora of the world, you will automatically save the physical environment."

— *Edward O. Wilson*

ABALONE

GREEN SEA TURTLE

CALIFORNIA BROWN PELICAN

AMERICAN WHITE PELICAN

CALIFORNIA LEAST TERN

ELEGANT TERN

GULL-BILLED TERN

LIGHT-FOOTED CLAPPER RAIL

BLACK OYSTERCATCHER

WESTERN SNOWY PLOVER

BLACK SKIMMER

LONG-BILLED CURLEW

OSPREY

PEREGRINE FALCON

NORTHERN HARRIER

BURROWING OWL

GRAY WHALE

SEA OTTER

Abalone
(*Haliotis sp.*)

Few mollusks can match the abalone when it comes to inspiring mankind's admiration and desire. Recognized as one of the most sought-after shellfish worldwide, the stately abalone (genus *Haliotis*) is the true King of the Sea. The beautiful shell of this slow-moving, snail-like mollusk can grow as large as a dinner plate. The abalone may be found in almost every tropical and temperate ocean in the world.

These aquatic warriors are protected by a thick shell that they use as armor to shield themselves from the dangers lurking beneath the water's surface. The abalone's shell grows with the creature after the larva metamorphoses into a juvenile. The difference between abalone and bivalve mollusks is that only the upper half of the abalone is protected by a shell, leaving the lower portion of the creature exposed. This allows the abalone's powerful foot and retracting abductor muscles to attach to rocks and other surfaces with amazing strength. On the surface of the shell are four to ten holes (depending on the species) that permit water circulation for respiration and waste release.

The common species of abalone that are found along the California and Baja California coasts are black (*Haliotis cracherodii*), flat, (*H. walallensis*), green (*H. fulgens*), pink (*H. corrugata*), pinto (*H. kamtschatkan*), red, (*H. rufescens*), threaded (*H. assimilis*) and white (*H. sorenseni*). The most common species in Southern California is generally considered to be the red. It and the black and green species are the most valued by humans. (High Tech High 2007)

The red abalone is found along the lower rocky intertidal zones and range as far as 180 meters offshore. In California, however, the species is generally found between six to 17 meters deep. This abalone is California's largest marine snail, growing 2.5 centimeters annually to reach almost 30 centimeters in size. Its shell is generally red or pink in color, but may vary depending on its diet; when the red abalone ingests red algae, its shell tends to become a deeper color.

When abalone reach sexual maturity they reproduce via broadcast spawning, an event in which a species disperses its sperm and eggs into the water current for fertilization. On average, a 20-centimeter abalone can produce more than 11 million eggs, though typically less than 1% of these survive more than a few weeks. The larval form of the abalone is known as a veliger, which lives off the yoke of its egg. This lecithotrophic feeding strategy means the embryonic and larval abalone are reliant on the nutritional reserves supplied by their mothers.

During its life, the abalone must avoid a number of predators, such as the sea otter and the sheephead fish. That is why it tends to be found hidden between or beneath deep rocks and crevices. As a juvenile, it feeds mostly on diatom films, but in later years, its diet often consists of large brown algae such as giant, bull, feather boa and elk kelp. The abalone has a unique method of feeding, catching drifting kelp with its foot, and then raising the food to its tooth-lined tongue, called a radula. (Morris, Abbott, and Haderlie 1980)

Abalone were once abundant along the coasts of California and Baja California. The abalone that lived along this coast were originally harvested by Native Americans as far back as 8,000 to 10,000 years ago. During low tide, coastal aborigines would forage and pry abalone from rocks using a stick. This did not dramatically affect the abalone population because the Native Americans only hunted the abalone in numbers that they needed for subsistence, which allowed the vast majority of abalone to be untouched.

Ironically, the first commercial slaughter of marine life along the coast of California had a positive effect on abalone. In a sense, the hunting of sea otters by Russian fur traders during the early 1800s most likely helped the abalone populations in the region, since the otter was a top predator and a keystone species within the kelp forest ecosystem.

DECLINE OF THE
Abalone

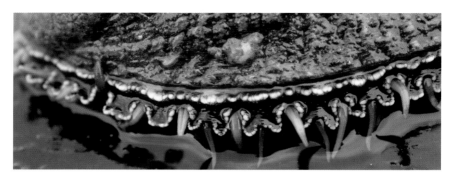

Tremendous hunting pressure on abalone and disease have led to the near-extinction of some species of this sluggish and tasty mollusk.

The low-level subsistence gathering of abalone that had been practiced by the earliest inhabitants along the coast changed dramatically during the 1860s when Chinese immigrants slowly started to migrate to San Diego after bad luck and racial tension during the Gold Rush. Since China placed a high value on abalone harvesting, these immigrants quickly took control of the industry in North America. They collected abalone by prying them off rocks with metal bars. Initially, this was done purely for the meat that was considered a delicacy in their homeland. Then they realized that there was a market for the creature's brilliantly colored, shiny shell; the Chinese seized the opportunity to not only hunt abalone for meat, but also to create beautiful jewelry and other items such as fishing lures from its shell. In 1879 more than $30,000 of cut abalone shell was being sold by the Chinese. This sealed the fate of abalone because of its great demand overseas. Soon, more and more people took part in its exploitation and the survival of this highly sought-after mollusk was threatened. (High Tech High 2007) (Rugh 2007)

Not long after the Chinese, Japanese-American divers tried their luck in the abalone market in the early

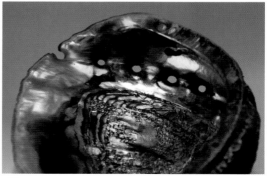

The brilliant iridescence of an abalone shell makes it a desired material for jewelry.

1900s. In 1935, the Japanese harvest peaked at 1.8 million kilograms of abalone off the California coast. They continued to hunt this animal until 1942, when the numbers captured decreased to around 74,000 kilograms as many Japanese American fishermen were moved to internment camps during World War II. (Leet, et al. 2001)

The species of abalone that were commercially fished included red, pink, green, black and white. Exploitation of abalone picked up again with the development of SCUBA technology in 1943 that allowed divers to breathe underwater without a cumbersome diving suit. The Aqua-Lung consisted of a valve-operated hose connecting the diver's mouth to a high-pressure cylinder worn on the back. For the first time, anyone in reasonable physical condition could explore the underwater environment for an extended period of time, unrestricted by the need to resurface for air. Also, people could dive for recreational as well as commercial purposes. SCUBA thus became the new threat to abalone and would continue to increase until legal actions were taken to finally protect its populations and allow it to regain the numbers lost by overfishing.

Although harvesting black abalone is now banned, disease threatens its survival. Withering syndrome results in severe atrophy of the animal's foot muscle. It is caused by a water-borne pathogen excreted in abalone feces and occurs in relatively warm water. The black abalone has become so rare it is now a candidate for protection under the federal Endangered Species Act.

RECOVERY OF THE
Abalone

Farm-raised juvenile abalone grown for restocking efforts in SPAWAR tanks.

After decades of exploitation, the number of remaining abalone became a cause for concern, and the California Department of Fish and Game decided the population could no longer support commercial harvest. Many abalone species had decreased to a point of unsustainability, so the collecting of abalone was halted in 1997. Additionally, only free diving (without the use of SCUBA gear) was permitted when hunting for abalone, limiting divers to the air in their lungs.

To restore populations of this unique mollusk, a variety of laws have been enacted to preserve abalone habitat and regulate recreational hunting. However, because the life history strategies and reproduction patterns of abalone are still not fully understood, sustainable management of the fishery is an ongoing process. The commercial fishery is closed and will likely remain so until some dramatic changes take place. (California Department of Fish and Game 1998; 2008) (High Tech High 2007)

Although rules were put in place to protect abalone, its value remains and poachers risk heavy fines to collect this mollusk. In the current market, each abalone can fetch $100. This means that poachers are more likely to ignore laws protecting abalone by deliberately taking more than the allowed three abalone per day.

As awareness and exploitation accounts increase, perhaps the public will be able to save this creature before it disappears from the coast forever. Current restrictions are respected by many and designate the size and number of red abalone that may be collected; only those animals whose shells are larger than seven inches, limited to three-per-day and 24-per-year bag limits This encourages the growth of the red abalone population. For other species, numbers are so low that breeding itself becomes a problem. Such is the case with the white abalone whose numbers are so low that its gametes (eggs and sperm), which are released into the ocean, often fail to encounter each other, preventing fertilization. To help this dying species, white abalone are being bred in captivity in land-based aquaculture environments. Only time will tell whether this effort will help save the once-abundant white.

Another method of recovery is being conducted by David Lapota with SPAWAR and the U.S. Navy in San Diego. His project involves breeding green abalone in tanks in Point Loma along the shore of San Diego Bay and then placing young tagged abalone on reefs within the Point Loma kelp bed. Recent restocking efforts from this project appear to be quite promising with 50% survival six months after introduction. Maybe there is chance for recovery of this imperiled species after all. Although Lapota speculates that it may be decades before the commercial abalone fishery returns, successful studies like his give hope that abalone may be able to recover. (High Tech High 2007)

Abalone populations are dependent on healthy kelp beds.

Green Sea Turtle
(Chelonia mydas)

The green sea turtle is a reptile that has inhabited the planet for nearly 150 million years and is found in oceans worldwide. Like other reptiles, it is cold blooded, acquiring its body heat from the surrounding environment. It has an outer shell made of hard bone that acts as camouflage by matching its environment—a protection from predators lurking above. The shell covers two major parts of its body: the dorsal (back) and ventral (stomach). The dorsal portion of the shell (or carapace) is covered with scales, with openings for the head, limbs and tail between the carapace and the ventral portion of the shell. (Hawxhurst 2001)

One of the unique aspects of the green sea turtle is that, unlike most turtles, it cannot retract its head inside the shell. The turtle surfaces for air approximately every 30 minutes, but when an adult turtle is asleep, it can stay underwater for up to two hours. This is because the adult can hold higher concentrations of carbon dioxide in its blood than most other air-breathing animals. Young turtles do not have this ability, so they sleep while floating on the water's surface. The turtle's body is adapted to the water by having fins that enable it to swim for great distances at an average speed of 56 kilometers per hour.

The turtle gets its name from the color of its body fat, which comes from the green algae that makes up its diet. The adult is a herbivore (plant eater), and depends on the bacteria in its stomach for healthy digestion of food. The juvenile turtle is carnivorous, with a diet consisting mainly of jellyfish and other invertebrates.

The adult sea turtle can weigh more than 225 kilograms. In San Diego Bay, the biggest sea turtle ever captured is a specimen named Wrinkle Butt, who weighs almost 250 kilograms. The lifespan of a green sea turtle is unknown, but it is possible it may live up to 250 years. The male and female turtle look alike until maturity, after which the male begins to acquire a long, thick tail, while the female retains a short, stubby one. Although the turtle lives most of its life in the ocean, it must surface for air and the adult female returns to land to lay eggs on sandy beaches above the high-tide line. These eggs take about two months to incubate, a period that leaves them vulnerable to predators. ("Marine Turtles" 2007) The eggs hatch at night, and the young turtles instinctively head directly towards the water's edge. ("Green turtle" 2008)

Because of the turtle's large size, an adult has only two main predators: sharks and people. However, the young turtle has other predators, such as crabs and birds that attack after it has hatched and as it crawls to the sea. It is the only sea turtle found in San Diego Bay, attracted here by warm waters generated by the power plant and large amounts of eelgrass, a favorite food. However, it is possible that sea turtles have been in the bay much earlier than the power plant. Indirect evidence of their habitation in the Pleistocene is suggested by fossil barnacle species that only reside on turtles found in the bay. (Rugh 2007) Today, it is estimated that 30 to 60 turtles thrive in the bay, which provides them with a protected foraging habitat while offering a prime study area for researchers. ("Environmental Services" 2007)

San Diego Bay is generally the turtles' northernmost dwelling habitat. However, when water temperatures are warm, they have been sighted as far north as southern Alaska. ("San Diego Bay Integrated Natural Resources Management Plan" 2000)

DECLINE OF THE
Green Sea Turtle

Almost half of the world's sea turtle species are currently facing the threat of extinction. Green sea turtles are endangered for a number of reasons: global warming; global pollution; coastal development, and poaching. One major international human threat is the harvesting of the turtle's eggs and the selling of its meat. This is popular in various regions, such as Southeast Asia, where turtle meat is considered a delicacy. Turtle parts are also used for jewelry; turtle shell combs and other shell products are popular on the black market. Turtle products are not only profitable but have cultural ties for many people. These practices continue despite laws that prohibit the buying and selling of the sea turtle in the United States and other countries with sea turtle habitat. (Stinson 2007)

In San Diego Bay the three main causes of endangerment are human interaction, commercial boating and pollution. Marine debris also has a huge impact on sea turtle survival. Nondegradable trash such as balloons, bottles and plastic bags are deadly to turtles; floating plastic products are mistaken for jellyfish, a part of the juvenile's diet. Plastics are extremely harmful because they are not easily digested and remain in the turtle's stomach for a long time and release toxic substances. Ingested plastics also can clog the turtle's digestive system, blocking the proper passage of food and causing starvation. Oil spills, chemical runoff and fertilizers disrupt the turtle's respiration, blood chemistry, skin and salt gland functions. Marine pollution kills the aquatic plant life it relies on for nourishment. There is now evidence suggesting that the disease fibropapilloma, which causes the growth of large, bulbous tumors on the soft tissues of the turtle, could be connected to ocean pollution. Once a sea turtle develops this disease, there is little likelihood of its recovery. The tumors often spread, ultimately killing it. (Hawxhurst 2001)

Green sea turtles can easily mistake plastic bags for natural food items. At the Chula Vista Nature Center, their diet is supplemented with lettuce.

Another leading cause of sea turtle death is accidental catch in fishing gear. Many countries, including the United States, have implemented laws to protect turtles and turtle nesting areas within their jurisdiction. The widespread use of turtle excluder devices (TEDs) by U.S. shrimp fishermen has reduced U.S.-caused turtle fishery-related deaths by as much as 97%. TEDs are "escape hatches" for turtles caught in shrimp trawl fishing nets, which allow them to get free and return to the surface to breathe. Unfortunately, the use of TEDs has not been adopted by shrimp trawlers worldwide, and more than 10,000 sea turtles drown each year. ("Green turtle" 2008) (Stinson 2007)

East Pacific green sea turtles nest along the Mexican coast, often returning to the same beaches where they were hatched and lay their eggs in the sand above the high-tide line. Each year beaches are lost to coastal development, leaving the female turtles without a familiar place to lay their eggs. Disturbances such as noise, lights and obstructions are disruptive to nesting areas and threaten this critical part of the sea turtle's life cycle. Without a safe and reliable place to lay its eggs, the turtle population will slowly dwindle to unsustainable levels. (Stinson 2007)

Studying the sea turtle is difficult because the animal spends the majority of its life at sea. Much is known about the land turtle and its nesting process, but little is known about what happens with the marine species. Determining the numbers of turtles is a challenge because generally only the female has been studied during nesting season. To make population estimates, scientists like Jeff Seminoff use complex algorithms and equations with information derived from nesting sites: how many eggs were unhatched or were crushed. Also as many turtles die young, it is difficult to establish how many survive long enough to reproduce. Scientists have to account for many factors in generating these estimates, making the process time consuming.

RECOVERY OF THE
Green Sea Turtle

The most effective protection plans for the sea turtle are national and international protection laws. In the United States, the green sea turtle is protected under the Federal Endangered Species Act, which prohibits hunting, injuring or harassing the turtle, or holding it in captivity without first obtaining a permit for research or educational purposes. Swimmers and divers should be aware that riding or touching a sea turtle is illegal as it causes the animal unnecessary stress. Fines for violating these laws may be as high as $100,000 and may include prison time. International law prohibits the trade of sea turtle parts or products under an agreement known as the Convention for International Trade of Endangered Species of Wild Fauna and Flora. Unfortunately, illicit trading of the sea turtle and its products still continues at an alarming rate. These laws need to be strictly enforced around the world and people made aware of the illegal trade of sea turtle parts.

The federal government plays a huge part in preserving the sea turtle. Humans can significantly improve the animal's habitat by remembering a few simple things. Reducing littering by putting trash in garbage cans and recycling helps to prevent plastic from getting into the oceans. Released balloons are popular events but are quite harmful to the turtle. The balloons eventually deflate and could land in the ocean. A juvenile turtle could mistake the de-

This green sea turtle is a new card-carrying member of the mark-and-recapture study coordinated by Peter Dutton of National Marine Fisheries Service. Photo: National Marine Fisheries Service Research Permit #1591

flated balloon for jellyfish—part of its diet—and die from ingesting the object. Another good idea is to wash a car on the lawn or at a carwash.

When people wash their cars on the sidewalk, dirty soap and oil go into the storm drains that flow into the ocean, which is harmful to all marine life. Education, along with raising awareness, is a powerful tool in the recovery of the sea turtle. The more people who know about such problems, the greater the likelihood of their involvement. (Stinson 2007)

In Mexico, the breeding population appears to be declining. As a result, an east Pacific green sea turtle recovery plan was prepared just for this stock. ("San Diego Bay Integrated Natural Resources Management Plan" 2000)

Peter Dutton of the National Marine Fisheries Service and the director of San Diego Bay Green Sea Turtle studies is a well-known turtle expert. He and other researchers and scientists are using technology advances to help the sea turtle. Satellite transmitters are attached to captured turtles to learn more about their lives and monitor their movements worldwide, which is valuable in preserving those areas identified as nesting beaches. A captured turtle is also given a number so it can be identified if caught again. Once a turtle is identified, it is weighed to determine its growth and get an estimate of its age. If the turtle originated from another country; scientists use the identification number to determine who was researching that specific turtle and provide that person with additional data. (Dutton 2007)

By learning more about the human effects on the sea turtle, we will be able to develop better laws for its protection and restoring its population numbers for future generations.

Biologists collect data before returning this sea turtle to the unnaturally warm waters of South San Diego Bay. Photo: National Marine Fisheries Service Research Permit #1591

Margie Stinson

As we walked towards the South Bay Power Plant, a chance encounter brought Margie Stinson, one of the first turtle scientists in San Diego, across our path. When she humbly introduced herself, we were shocked by our luck and jumped at the prospect of interviewing her. Standing in front of the looming power plant, there was no better environment to describe the resilience of the green sea turtle to survive in such urbanized conditions.

Margie Stinson is one of the first to study sea turtles in San Diego Bay, beginning her journey as a grad student at San Diego State University. It was during her studies that she identified the presence of sea turtles in the waters warmed by the power plant. Now, as assistant professor of marine biology at Southwestern College in Chula Vista, she continues to study and be an advocate for the turtles that come to the thermally altered shores of the South Bay.

Student Researcher (SR): How did you first get interested in researching sea turtles?

Margie Stinson: When I was an undergraduate student at San Diego State in 1976, I didn't know I was going to end up working with sea turtles. I thought I would be working with elephant seals, dolphins, blue whales or gray whales because I was spending so much time in Baja California working with those animals. But every time I would talk to my major professors about an idea I would have for my thesis, they would say, "Oh, no." This happened about five or six times. So, I went to speak with Captain Eddie McQueen of the *Pacific Queen*, and I told him about what I learned in school that day, which happened to be about green sea turtles migrating in the Atlantic from the shores of Venezuela and Brazil. After learning about my interest, he and his crew brought me down to the south edge of San Diego Bay to observe sea turtles first hand. On the south shore of San Diego Bay, we lay in the mud and the rain with our cameras and photographed what I thought were going to be ducks, but which were actually the heads of turtles.

With those pictures, I went to Scripps [Institution of Oceanography] to show them to Dr. Carl Hubbs. He was shocked to see the presence of turtles. He then asked me what I wanted to do with my findings. I said I wanted to learn everything I could about this turtle and any others in San Diego Bay. With that he opened his file drawer and started handing me file after file of pictures, data sheets and field notes that he had kept his whole lifetime, some stretching back to the 1800s. Well, I brought this information to my major professors and they could not believe that there were sea turtles in San Diego Bay. I went through all the scientific literature to find every article ever written about sea turtles anywhere in the world, in any language and in doing so, I found that there were only a handful of articles on the sea turtles in California.

I started making a list of all the questions I had about sea turtles in San Diego Bay. When I took this list to my major professors, they were so excited they didn't attempt to limit me. They wanted me to find the information on every question and, after almost a decade, I finished the initial phase of research and ended up with a 600-page master's thesis.

SR: What research did you do after your master's thesis?
Stinson: Through my research I found out that turtles appeared to be a seasonal creature in San Diego Bay. After learning this, I wanted to find a way to capture them and put on some kind of device to track their migration. This was during the early years of telemetry, and I was fortunate enough to work with the U.S. Navy here in San Diego, and they helped me design a telemetry radio. It took about two years of trial-and-error to learn how to successfully capture a sea turtle. This was a first for that type of project and most of what we did was prototypic.

Peter Dutton's associates from the National Marine Fisheries Service monitor the growth and movement of turtles in the bay. Photo: National Marine Fisheries Service Research Permit #1591

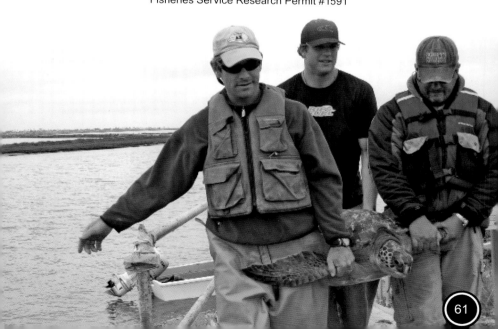

SR: How do you attach the tracking device to the sea turtles? And what have you learned from tracking them?

Stinson: We superglued the radios onto the the dorsal shell and released the turtle. I also used a tiny radio about the size of a big vitamin pill that could be swallowed. Both radios could measure the temperature of the turtle's environment. From these I was able to compare water and internal body temperatures until the turtle would defecate

the radio out. I didn't know whether it would work and if it would be a few hours or a few weeks, but it turned out the turtle would keep its radio for up to two weeks. This meant I would have several weeks of tracking the turtle and collecting both water temperatures and internal temperatures to find out if the reptile was thermo-regulating like other reptiles.

SR: What do you believe has caused the decline of sea turtles both locally and worldwide over the years?

The location where we interviewed Margie Stinson could not have been better for portraying the unnatural environment of the San Diego sea turtles.

Stinson: Many of the populations of sea turtles are suffering from a virus called papilloma. It starts out as a skin wart and then it can go systemic. Some populations may have 50% of its animals infected with this virus, and some go as high as 80%, and it can be fatal. Now, researchers are finding a connection between the virus and pollution. Sea turtle populations near urban, highly polluted areas are those you would predict to be most at risk for getting this virus.

SR: Has global warming affected the places where sea turtles migrate?
Stinson: Sea turtles are a tropical and subtropical species. As the isotherms, or sea-surface temperatures, change globally you may see a global shift in the location of populations. In my studies, not only did I work with the green sea turtle in San Diego Bay, but other species of the turtle found along the coast of Baja California. I looked at sea-surface temperature data that stretched back to the early 1900s for every oceanographic collecting station, such as Scripps Pier. I figured out if the temperatures were abnormally high (El Niño), abnormally cold (La Niña), or normal for that particular location for every year. I had 21 of these stations along the coast. Then for every sea turtle sighting that I was able to get, I triangulated and determined if that sighting occurred when there were El Niño, La Niña, or normal [ocean temperatures]. What I found was that along the coast of California, we have a sea turtle season regardless of water temperatures.

California Brown Pelican
(*Pelecanus occidentalis californicus*)

Many California observers can relate well to the California brown pelican, (a local brown pelican subspecies,) since it spends much of its time surfing waves, fishing and lounging on the shore. The brown pelican is one of the most distinctive seabirds, with its food-capturing pouch, graceful soaring flight and plunging dives. It is a common sight along the shores and beaches of San Diego Bay where it may be seen flying up and down the coast or seeking a meal from the bay's waters.

The California brown pelican is the smallest of the *Pelicandae* family. Mature adults are usually 106–138 centimeters in length with a 200–254 centimeter wingspan, a large brown body, a yellowish bill and short, dark brown legs and feet. The adult birds have whitish heads and distinctive blue eyes. During mating season, the male develops attention-attracting colors, such as a red patch on its head and a slight yellowing in the head plumage. However, the most unique attribute of the pelican is undoubtedly its food pouch, fittingly described by Thomas Pearson, (who helped found the National Audubon Society), with the phrase, "… it is doubtful whether a pelican could fly at all with his burden so out of trim." (Pearson 1936)

The California brown pelican spends most of its time in saltwater, frequenting stillwater bays and less turbulent seas. Unlike its close relative, the American white pelican, this open-water bird is a true mariner, depending on the ocean for food.

The bird's daily diet consists of three to four pounds of herring and anchovy-like fish. Its food-catching plunge is a remarkable blend of speed and accuracy; from heights as great as 20 meters above the water's surface, the pelican pinpoints schools of fish, plunges into the ocean, gracefully scoops its prey from the water, and then thrusts its head back to swallow the meal whole. As part of the process, the bird unfolds its tucked wings just as it touches the water's surface. This instinctive move prevents it from plunging too deep into the water (Anderson, Gress, and Mais 1982).

The brown pelican builds nests of sticks on the ground, typically on islands or offshore rocks. March is the peak egg-laying month, with an average of three eggs per clutch. The young fledge in early summer. The food exchanges that occur between parents and helpless chicks are some of the more bizarre in the avian world. At first, the parents deposit a regurgitated fishy soup into the front end of the chick's pouch, but when the chick is about half-grown it hunts for meals in the inner reaches of its parent's gaping mouth, its entire head and neck disappearing into this abyss. (Pearson 1936)

The California brown pelican breeds primarily on islands in the Sea of Cortez and along the West Coast of Baja California. A considerable amount of research has concentrated on the bird's population in North America due to the decline of nesting populations in the documented northern colonies. The only breeding population in U.S. waters is the Southern California Bight population, which consists of breeding birds on the Channel Islands and several islands off Baja California. Locally, the birds frequent San Diego Bay and San Diego coasts during pre- and post-breeding season to feed and socialize. Like other seabirds such as cormorants, shearwaters and storm-petrels they are highly gregarious. Today, the California brown pelican can be seen along the U.S. West Coast. (Briggs, et al. 1981) (Unitt 2004) (Burkett, Logsdon, and Fien 2007) (Peterson 1934)

While traveling up and down the California Coast and the Baja Peninsula, American writer John Steinbeck became highly familiar with the movements of the pelican: "A squadron of pelicans crossed our bow, flying low to the waves and acting like a train of pelicans tied together, activated by one nervous system. For they flapped their powerful wings in unison, coasted in unison. It seemed that they tipped a wavetop with their wings now and then, and certainly they flew in the troughs of the waves to save themselves from the wind. They did not look around or change direction. Pelicans seem always to know exactly where they are going." (Steinbeck, Ricketts, and Astro 1995)

DECLINE OF THE
California Brown Pelican

Since the turn of the twentieth century, brown pelican populations have faced one battle for survival after another, all of which have significantly damaged the bird's population. During the late nineteenth and early twentieth centuries, pelican feathers were used by milliners to decorate hats, and when food shortages occurred during World War II, its eggs were collected for human consumption. The pelican was also killed because fishermen believed it threatened commercial fish resources.

The California brown pelican population was pushed near extinction with the introduction of the infamous insecticide, DDT. Once ingested, this pesticide latched onto a pelican's fatty tissue and accumulated in its system. In one case, the DDT concentration in a pelican was measured at nearly 100 parts per million. DDT was also found to affect a pelican's hormones and reproductive cycles, which involved the mobilization of calcium from female bones. For more information on DDT, see "An Inside Look: DDT in the Environment" on page 76 of this book. (Small 1974)

The dramatic drop in brown pelican numbers quickly drew the attention of biologists and federal agencies alike, and in June 1970 it joined many other seabirds on the Endangered Species list.

Human interference is another leading cause for the drop in the bird's population. Careless fishermen pose a threat to the pelican by leaving fishing lines and hooks in feeding and roosting areas. The bird is attracted to the struggling fish and may get tangled in lines and strangle or starve to death. Frequently, fishing hooks puncture the pelican's pouch, leaving large holes that prevent effective hunting. Humans further endanger the bird by disturbing its nesting areas. Tourists, researchers and photographers who get too close often chase away the parents, leaving the chicks alone and vulnerable.

In 2001, the California brown pelican faced a new threat when avian botulism began to show up in many bird species, including the pelican. The disease attacks water birds and is caused by a toxin produced by anaerobic bacteria. The bacterial disease first infects fish and is then spread to birds that ingest the toxic fish. The outbreak of this disease, reported in 1996, resulted in the death of more than 1,000 California brown pelicans and 4,400 other seabirds in the Salton Sea (Beacham, Castronova, and Sessine 2001). Although a direct link has not yet been demonstrated conclusively by scientists, the spread of avian botulism could pose a major threat to the recovery and preservation of the already fragile pelican population.

Disease, destruction of nesting stock and human interference contributed to the dramatic decline of the brown pelican population. Yet even without these threats, the species faces an uphill battle to survive because of natural forces. Similar to other seabirds along the coast of the Eastern Pacific, variations in the estimated population numbers of the California brown pelican seem to be related to the natural cycles of El Niño-Southern Oscillation (ENSO) phenomena. During the ENSO period, breeding populations have been greatly reduced in number and some traditional breeding grounds have shown no nesting. These periods, which occur on roughly five- to seven-year cycles, result in major shifts in currents, water temperature, nutrient levels and fish stocks for diving birds. (Anderson, Gress, and Mais 1982)

The California brown pelican has made an amazing recovery from DDT, and now must continue to adapt to a heavily developed coastline.

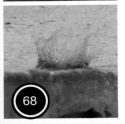

RECOVERY OF THE
California Brown Pelican

Since the ban of DDT, the California brown pelican has made a remarkable comeback. The majestic and resilient brown pelican has steadily rebounded from the low population numbers it sustained during the 1960s and 1970s. While this book was in progress, the California brown pelican was removed from the endangered category under the Endangered Species Act, a great success for this iconic bird. However, the enthusiastic reaction towards the rebounding brown pelican population may be somewhat overzealous and presumptive. The recent endangerment of the brown pelican is not something that should be readily forgotten.

Additionally, the Eastern brown pelican (*Pelecanus occidentalis carolinensis*) has also been removed from the federal list of endangered species. Much of this recovery is due to Rachel Carson's legacy, but also a multitude of nonprofit environmental groups that have emphasized educating people about the pelican's plight, minimizing contact at nesting sites, and attempting to stop land development near or on pelican-nesting sites. (Beacham, Castronova, and Sessine 2001) In 2006, it was estimated that there were 70,680 breeding pairs of the pelican. (Anderson, Gress, and Mais 1982) With strong stewardship and awareness campaigns this recovery will hopefully continue.

A California brown pelican dives for fish.

WHAT MAKES THE SEA?

What makes the sea?
Is it the turbulent murmur of the frothy waters?
Or the buttery softness of sun-bleached sand?
No, it is the ambassador of the waves:
The elegant pelican.

Be they white or be they brown,
Wherever the sea meets land,
The pelican can be found.

From high above, at the break of day
Pelicans soar across the bay.
Then suddenly with a flurry of wings
Pelicans dive with the authority of kings.

What makes the sea? A pelican.
What makes a pelican?
That is a story to be told by another man.

— Rachel Bouffard

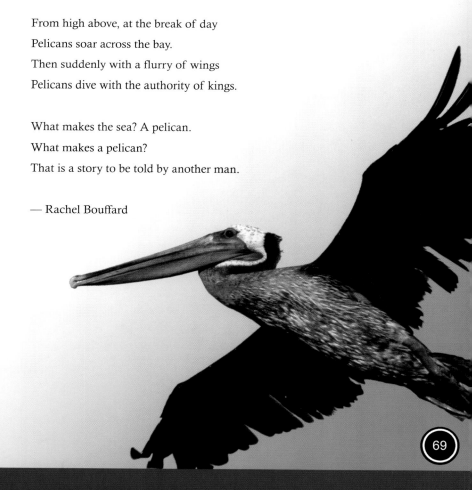

American White Pelican

(Pelecanus erythrorhynchos)

The American white pelican is an artifact of an ancient time. Its ancestry dates to the Miocene era, and its thick, white eggshells remind us of its reptilian lineage. John James Audubon named the American white pelican. "I have honoured it with the name of my beloved country, over the mighty streams of which, may this splendid bird wander free and unmolested to the most distant times, as it has already done from the misty ages of unknown antiquity." (Audubon 1842)

Like the California brown pelican, one of the American white's most notable features is its large gular (pouch) enabling it to capture a wide range of fish. The American white is one of the largest in the Pelicandae family. In length, its body ranges from 1.2 to almost 1.8 meters, while its wingspan can reach more than 2.7 meters. An air-filled skeletal structure and air sacs provide the pelican with maximum buoyancy. The white pelican possesses a long, yellow bill with pouch as well as short, yellow legs and feet. The adult displays odd bill protuberances during the breeding season. Another notable difference is its tendency to tuck its head and neck to the side of the body when nesting or sleeping. Its unmistakable white feathers and black wingtips make it somewhat similar in appearance to that of a whooping crane. (Udvardy and Rayfield 1977) (Shuford, 2005)

The American white pelican frequents freshwater habitats and has developed special methods for adapting to varied ecosystems. In contrast to the fishing technique of the California brown, the American white does not dive for its food. Instead, it employs group fishing tactics—schools of fish are herded into shallow water so they can then be easily collected near the surface. Additionally, the American white has the ability to shift fishing strategies to optimize feeding efforts in lakes, bays and streams depending on the availability of prey species. On average, this pelican consumes nearly two kilograms of fish daily, including carp, perch, catfish, jackfish, shiners and other species that are found in San Diego's bodies of freshwater and estuaries. (Peterson 1934)

The American white pelican tends to nest in secluded areas. During mating season it gathers with other white pelicans, after which the group separates to nest, often on the grounds of small islands, away from the dangers of terrestrial mammals. Each white pelican clutch consists of up to two eggs, which usually fledge during mid-summer. The white pelican does not breed in San Diego; however, it takes shelter in San Diego's estuaries to feed and recuperate for extended periods of time. Local populations of the American white pelican are relatively small because of extensive estuary draining for land development. However, in recent years, some of the white pelican population has relocated to San Diego due to fish declines in the Salton Sea, and this has accounted for a local increase in the bird's numbers. It is most commonly observed at Lake Henshaw at the base of Mount Palomar or in the southern-most reaches of San Diego Bay.

DECLINE OF THE
American White Pelican

American white pelicans are a social species. Feeding, for example, is a group activity.

Prior to the 1960s, the future of the American white pelican appeared dim, as the effects of human interference, the increasing presence of contaminants, changing water levels, disturbance from construction and noise pollution all contributed to colony destruction and loss of habitat—the two leading causes of the bird's decline. Until the mid-twentieth century, the white pelican population decreased drastically with increases in human expansion and development throughout the country. Rivers were diverted and land was reclaimed for agricultural and metropolitan development. It has been speculated that more than 90% of California's wetland habitat has been lost. Rivers and flooding patterns that once maintained large wetlands no longer exist because of negative human intervention.

Nesting sites and foraging areas for the white pelican have been reduced considerably, shifting concentrations of the bird's population to small, dense areas. These population clusters are especially vulnerable to drastic environmental changes such as those caused by famine, poisoning or human disturbance. Additionally, the white pelican is extremely sensitive to change and will abandon its nest and young if frightened. (Shuford 2005) (Murphy 2005)

Wildlife poaching is a recurring theme related to the loss of the pelican in recent history. Humans often viewed the bird, which consumed large amounts of fish each day, as competition for overexploited fish resources. Consequently, humans kept the pelican population minimized to guarantee maximum fish yield for themselves. However, the number of American white pelican killings dropped after the bird was granted protection under the federal Endangered Species Act.

Another contributor to the population decline of the white pelican is the emergence of avian botulism, caused by a potent bacterial toxin. This illness causes muscle weakness or paralysis in affected birds. In 1996, more than 15,000 birds of various species died in a severe outbreak of botulism at the Salton Sea in California, one of the few remaining pelican breeding sites in the state. American whites constituted more than 50% of the birds that died in this disaster—the largest death toll ever recorded for the species. Since then, smaller-scale population declines have occurred, but another severe avian botulism outbreak remains a possibility (Rocke, et al. 2005)

As with the California brown pelican, the white pelican population in California and several southern states was decimated by the unrestricted application of DDT before its usage was finally banned in the United States in 1972. Affected population clusters studied in the early 1960s displayed similar symptoms—loss of coordination, tremors, convulsions, and fewer young birds because of increased egg fragility. For more information on DDT, see "An Inside Look: DDT in the Environment" on page 76.

RECOVERY OF THE
American White Pelican

Although the white pelican was initially protected by the Migratory Bird Treaty Act of 1918, the true recovery of this bird's population did not begin until the 1960s when improved protective legislation, conservation efforts and public awareness contributed to its revitalization.

White pelican breeding surveys conducted between 1979 and 1981 in North America estimated more than 109,000 breeding individuals in 55 colonies. In contrast, available data from surveys conducted between 1998 and 2001 for 42 colonies in North America estimated a total of 134,000 breeding pelicans, an increase of about 2,200 birds per colony. (Anderson and King 2005)

Today, the California Department of Fish and Game (CDFG) faces a great challenge. Once pelican populations are re-established in California, the CDFG hopes to further promote the bird's presence within the state. Some recommendations include increased habitat maintenance, protection of breeding sanctuaries along California's coastline, the creation of artificial nesting islands at former breeding areas to encourage recolonization, and a continuation of monitoring population trends to ensure reproductive success for all colonies.

An Inside Look:
DDT in the Environment

Of all mankind's creations of the nineteenth and twentieth centuries, Dichloro-Diphenyl-Trichloroethane (DDT) has been one of the most devastating to the environment. It was created in 1874 by German chemist Othmar Zeidler. Subsequently, Paul Hermann Müller, the Swiss scientist who in 1939 first recognized its insecticidal properties, won the Nobel Prize for his discovery. In the following decades, DDT was used extensively throughout the world, and was hailed as a miracle method for controlling the spread of malaria and other diseases commonly carried by insects.

The danger of DDT lies largely in two of the toxin's properties: bioaccumulation and biomagnification. Bioaccumulation is the process by which organisms amass toxic substances at a rate that is greater than that lost. For example, a human who regularly eats food processed from grains that have been treated with DDT will accumulate gradually greater quantities of DDT, which take an extremely long time to dissipate (DDT has a half life as long as 15 years). Bioaccumulation works frighteningly in combination with biomagnification—the process by which species higher on the food chain accumulate higher levels of a substance than species lower on the food chain. For example, if grains treated with DDT are in cattle feed, cattle will build up even more DDT than the grains did, because they are being fed large quantities of the treated grains. By the time a human eats meat or drinks milk from the same cattle, that person is absorbing an extremely unsafe quantity of DDT. An individual storing DDT even as low as three parts per million is considered to be more susceptible to serious heart problems, and humans with no known exposure to DDT have been found to have an average of 5.3 to 7.4 parts per million. It is easy to see how severe a threat DDT is to both the health of humans and nonhumans. (Carson 2002)

This toxic pesticide travels with surprising speed throughout ecosystems. The pelican, one of the animals that has been most affected by DDT, serves as an example: runoff from fields that have been sprayed or dusted eventually makes its way to the ocean, where it is absorbed by phytoplankton (tiny marine plants). Zooplankton (tiny marine animals) feed on the phytoplankton, which are in turn eaten by anchovies. The California brown pelican eats a large number of small fish such as anchovies, so may have stored as much as 100 parts per million. This means that the concentration of DDT is approximately one million times greater than its initial state when it was sprayed on agricultural crops far from the ocean. (Small 1974)

The pesticide has also been interfering with the hormone cycle and production involved in the mobilization of calcium from female bones. In many birds, this inhibits the ability to produce a strong eggshell. As a result, some bird eggs are as much as 20% thinner than normal, and parents attempting to incubate these eggs will accidentally crush them. In 1969, this was recognized as a definite problem in the California brown pelican when nesting production virtually ceased. DDT has many other effects on wildlife and humans that have not been addressed here, and often harm different species in unique and unexpected ways. Species affected by DDT that are highlighted in this book include the California brown and the American white pelican, the osprey, peregrine falcon and northern harrier. (Small 1974)

Although it was the 1962 release of naturalist Rachel Carson's groundbreaking book, *Silent Spring* that exposed the unprecedented destruction DDT had caused the environment, wildlife and humanity, there had been concerns regarding DDT in the scientific community since the 1940s. *Silent Spring* was the first time the dangerous chemical was subjected to any serious criticism. The book had a tremendous impact on the public view of DDT and, in 1972, the use of the synthetic pesticide was banned in the United States. However, despite this, DDT is still used in many parts of the world.

Fish-eating birds like pelicans were critically impacted by DDT due to the biological magnification of DDT that occurred at each step of the food chain.

California Least Tern

(Sternula antillarum browni)

The California least tern is the smallest of terns, yet it received considerable recognition when it was listed as endangered in 1970 by the U.S. Fish and Wildlife Service (USFWS). Only 500 pairs of this graceful and beautiful tern were the estimated number at that time. The least tern is also a conservation success story as careful protection has increased the number of breeding pairs to 7,000 as of 2006. (Marschalek 2007)

Visually, the least tern is most notable for its white underparts, dark gray back, and wings that have black outer primary feathers. It is about 22 centimeters in length, with a wingspan of 50 centimeters; its small size makes it one of the most recognizable of terns. The head and bill are distinctive; a black cap with no crest dons the tern's head, and a white forehead patch goes from just behind the eyes down to its pointed yellow and black bill. All of the least tern's body is solid in color and texture. If one is lucky enough to observe the bird up close, its dark brown eyes will also be apparent. ("Least Tern *Sterna antillarum*" 2006)

The least tern commonly nests in open areas to increase predator visibility. Nesting areas include shelled beaches or open sand or gravel areas near the shore above high-tide lines. It is not as common, but certainly not unheard of, for a least tern to nest inland on river sandbars. It enjoys areas that are free from disturbances such as loud and unusual noises and human and domestic animal activity. Recently, in San Diego County, the tern was seen in large numbers at the Santa Margarita River mouth, the Naval Amphibious Base Coronado, Batiquitos Lagoon, Tijuana Estuary and Lindbergh Field. (Marschalek 2007)

The tern eats almost any small fish it can find in open areas of shallow water. Some of the places in which the bird has been known to feed are bays, ocean shore, tidal ponds, salt marshes and river outlets close to its home. Its aerial acrobatics feature torpedo-like dives from above, and also a skimming action along the water's surface. Small fish are no competition to the graceful yet deadly dives of the least tern.

Major nesting and breeding grounds for the California least tern are located along the California coast. The bird arrives from southern, warmer climes between April and May and builds its nest within two to three weeks after arrival. The male tern establishes and defends its territory vigorously, dive-bombing any intruders to scare them off. During the course of this study, this activity led to student researchers having their cameras, shirts and binoculars decorated with guano while monitoring tern nests with local biologists and refuge coordinators.

After the bird finds its mate, both individuals select a sandy area. The female will lay clutches of usually two eggs, but sometimes one or three eggs. Both male and female terns incubate and defend the eggs from 21 to 28 days; and once the eggs are hatched, both parents feed and brood with the chicks. The chicks leave the nest not long after hatching, and often wander as far off as 180 meters away. The young fledge after 21 days and leave the colony after two to three weeks. After early September, it is rare to find a least tern in San Diego.

DECLINE OF THE
California Least Tern

Extensive human interference in the life of the California least tern originally caused the species to plummet towards extinction. It is improbable that the beachgoers of Southern California will ever give back their large stretches of sand that they have taken away from the bird. Although its numbers are recovering, the least tern's safety is being continually threatened by encroaching human populations.

The least tern population dropped significantly for many reasons, most of which can be linked back to human infiltration of the species' primary habitat. Between hunting the bird for its feathers, clearing and developing its natural nesting areas, and dredging the bay, humans have been the cause of much of the damage associated with the bird's decline in and around San Diego Bay.

Abundant prior to the late nineteenth century, the least tern was on the brink of extinction primarily because of hunters seeking its plumes and skin. Much like other members of the Sterna family, its feathers were used to decorate hats during the days of the thriving millinery trade. By the turn of the century, about 100,000 terns were killed annually for the trade, resulting in a dramatic decrease in its numbers. Protective legislation in 1918 allowed the species to recover in the early twentieth century. In recent years, however, new pressures have caused a decline in the least tern population. (U.S. Fish and Wildlife Service 1918)

Modern development began with the dredging of internal areas of San Diego Bay, thus destroying much of the existing mud flats and sandy beaches where the least tern makes its home. Since then, land use around the bay has expanded and resulted in few

A headdress decorated with tern feathers.

areas of undeveloped habitat reserved for nesting birds. Areas that used to be popular nesting sites back in the mid-twentieth century have now been paved over to accommodate local businesses or residential housing. Because of this, many terns have had to become increasingly creative in choosing nesting areas—from abandoned lots to the tarmac at the San Diego International Airport (Lindbergh Field). Besides development, the least tern will often abandon nesting sites if vegetation or other debris overruns the area. It is important for the tern to nest in a cleared area so that it can watch its nest from above, in case a predator shows up. If popular breeding areas are not cleaned annually, the least tern is likely to abandon the site and move elsewhere. (Patton 2007)

Along with development and hunting, the least tern faces another big threat because of its many animal predators. Since the bird is a ground nester, it is highly vulnerable to destruction by many animals. Both feral cats and rats make meals of least terns and their young, a sad situation that has no simple solution—the feral cat population is rampant and there is no way to protect the least tern from rats. In addition, other predators include foxes, coyotes, raccoons, crows and ants. During the 2004 season, the CDFG reported the main predators were coyotes and American crows—these two species alone accounting for as many as 676 and 1,022 least tern deaths, respectively. Ironically, two of the imperiled species in this book are on the "black list" for the California least tern— the peregrine falcon and the gull-billed tern. When a peregrine does enter a nesting area, the adult terns will rise en masse, come together as a single

Billy Stewart designed this catch-and-release trap for relocating predatory species in least tern nesting sites.

flock and spiral up into the sky. A gull-billed tern was observed during the course of this study preying on a least tern chick at the Tijuana Estuary river mouth, flying off with the chick 100 meters to feed its own fledgling. Similar predatory events have been documented in other nesting areas of San Diego. (Unitt 2004) (Patton 2007)

Other natural factors that the tern faces include storms or environmental anomalies that result in lack of food. Storms have been known to kill many young terns. As nesting areas have decreased, the effect of astronomical high tides on tern eggs has also had an impact on the bird's population. Additionally, with recent global warming trends, local fish populations are being driven farther and farther away from least tern nesting sites. This results in little or no food for the young chicks. There is evidence that suggests the tern population may be shifting its nesting areas to the north to combat the issues it faces because of global warming.

The decline in the least tern population can mainly be attributed to the destruction mankind has imposed on the species. The bird was on the brink of extinction because of the devastation caused by humans. However, it has regained some ground and is beginning to make a recovery. This does not mean that it is not in danger. The California least tern is still threatened by many human and environmental pressures, and it will take the continued efforts of many dedicated people to save both this beautiful bird and its imperiled habitat.

An Inside Look:
LEAST TERNS AT THE AIRPORT

Tourists coming to San Diego during spring and summer and using the San Diego International Airport (Lindbergh Field) may have the best vantage point for viewing nesting behavior of the California least tern. It is indeed strange and remarkable to see the little white birds alongside a runway by an enormous aircraft. With the limited number of nesting sites throughout San Diego, some least terns have adapted to the bustling life of the airport, improvising new breeding grounds.

Between the months of April and September, as many as 90 terns can be seen at Lindbergh Field. The airport's tarmac suits the bird's nesting needs because it mimics the open, flat beaches on which it would typically nest in natural conditions. Over the years, Lindbergh Field has become the nation's busiest single-runway airport, with a flight leaving every 90 seconds. To protect the least tern, volunteers place small, brown shrubs and plastic fencing to isolate it and its nesting sites. Volunteers under the guidance of Robert Patton and Richard Gilb also band newly hatched chicks. In 2006, there were more than 122 nests and 55 new chicks spotted on airport property. This story perfectly illustrates nature's resilience in the face of human pressure; one of San Diego's most treasured avian species has triumphed even under the most dubious of circumstances.

RECOVERY OF THE CALIFORNIA LEAST TERN

Since its addition to the federal list of endangered species in 1970, the California least tern has been the subject of extensive management programs that aim to revitalize its population. The bird is subject to a larger monitoring and protection program than the other terns because of its low numbers. In San Diego numerous conservation groups and government agencies have collaborated and created various programs that deal with the recovery of the bird through nesting-site renovation and other activities dealing with the species' breeding grounds. These groups include the Port of San Diego, San Diego Audubon Society, the USFWS, the CDFG, the U.S. Navy, the U.S. Marine Corps and the San Diego Regional Airport Authority.

The San Diego Audubon Society has become one of the leading stewards in least tern recovery. The group and its passionate and concerned members have contributed countless hours and funds for the bird. When it is not breeding season, the society and its community volunteers spend an average of 500 hours maintaining least tern nesting sites in Mission Bay and around San Diego. The work primarily consists of removing vegetation and other debris from the bird's nesting areas that might deter it during the breeding season. The San Diego Audubon Society has received a variety of grants used specifically for least tern recovery. With these grants, members of the society were able to install improved fences that enhanced protection for the chicks. In addition to maintaining its habitat, the Audubon Society has an education program that works with students to create least tern decoys. These decoys are placed in protected nesting areas throughout San Diego, such as North Fiesta Island in Mission Bay, to promote safe breeding in places where local agencies can monitor

the bird's progress. ("California Least Tern IBA/Habitat and Education Project" 2005) ("SDAS Conservation Projects" 2007)

Although the U.S. Navy's decades of dredging and geographic modification have displaced many species throughout San Diego Bay, such as the least tern and elegant tern, it has begun to collaborate with a variety of other agencies, like the USFWS and CDFG, to protect terns on government property. Several designated areas on Naval Amphibious Base Coronado and the Naval Air Station North Island are closed off during the tern's breeding season.

CALIFORNIA LEAST TERN
NESTING AREA
DO NOT DISTURB

THIS ENDANGERED BIRD IS PROTECTED UNDER CALIFORNIA STATE LAW AND FEDERAL LAW.

DEPARTMENT OF THE INTERIOR
U.S. Fish and Wildlife Service
Endangered Species Act of 1973
Public Law 93-205

State of California
THE RESOURCES AGENCY
Department of Fish and Game
Fish & Game Code, Sections 2050-2055
and Title 14, CAC, Section 670.5

In 2002, the Navy began a program where tern eggs that were laid in undesignated areas were collected and raised in captivity. Out of the 50 eggs collected that year, there were 43 hatches and 31 chicks survived to reach their fledgling stage. All but six were released back into the wild. (Burr and Conkle 2003)

Stewards are also getting involved in least tern recovery through predator control and tern banding. Certain species of birds and mammals that present a threat to the tern's nest, eggs and chicks are managed by both lethal and nonlethal means. Species that are federally or state listed or of special concern are relocated to areas where they cannot harm the terns. Tern banding consists of banding the legs of all the chicks that can be captured or collected shortly after hatching. Two types of bands are placed on each chick, a USFWS and site-specific color band. These help in identifying, recording and tracking the terns throughout San Diego County. (Burr and Conkle 2003)

There are additional protected least tern nesting sites in San Diego. The most popular are those in Mission Bay, Mariners Point,

Robert Patton, a local biologist and tern expert, bands recently hatched least terns.

the Salt Works, Silver Strand and north Fiesta Island. These sites are constantly being changed, renovated, cleared and improvised to make more optimal breeding sites for least terns.

Over the years, these programs have benefited the tern population. In 2006, an estimated 7,000 breeding pairs were spotted in San Diego; however, only 2,500 fledgings survived that year. Although the least tern breeding success is still low, the USFWS is now considering downgrading the tern's status from "endangered" to "threatened."

According to Robert Patton, a leading least tern researcher who has been monitoring the bird's nesting sites around San Diego Bay since 1981, if the tern is downgraded, the population would likely crash within the next 15 years. Although the bird is slowly repopulating, it is only due to the strong efforts being exerted to save them. If it were to be taken off the Endangered Species list, all efforts to protect them would deteriorate because the population would begin to fall faster than it could be sustained.

To protect and restore the least tern population, it needs to remain on the Endangered Species list. Additionally, the hard work of valuable community stewards must continue to be supported. (Patton 2007)

Chris Redfern

As we stood and looked out over the expanse that is a rounded jetty protruding into Mission Bay, Chris Redfern, San Diego Audubon Society Executive Director, pointed to the sky and told us to watch carefully. The light and agile California least terns were darting and diving, gracefully attacking a large and cumbersome sea gull, which had the misfortune to fly over their well-established nesting site. We were soon to discover that the protective nature of the tern parents was paralleled in Mr. Redfern's commitment to the San Diego Audubon Society, where he works tirelessly to protect and revitalize the coastal habitat in and around San Diego Bay. He is a leader who not only organizes volunteers for events but whole-heartedly assists in restoration and cleanup projects. His commitment and dedication to saving local habitat and its natural avian inhabitants is obvious in his vast knowledge of the many efforts to protect the least tern. As we stood by him, listening attentively, each of us was grateful for his contribution to animals whose space we share. With the high-pitched squawking cry of the terns in the air, the sun washing down on us with its warm light, we stood on the surrounding sand dunes to listen to Mr. Redfern's knowledge and recovery stories related to this endangered bird.

Student Researcher (SR): Why are California least terns important to San Diego Bay?

Chris Redfern: It's difficult to pinpoint why a specific species is important to the bay. One of the things that makes the least tern important is that it is an example of how we manage endangered species. Back in the early 1970s, before the Endangered Species Act, there were only about 500 or 600 pairs of this little bird nesting along the California coast. Since they were declared an endangered species, we have been able to protect them by providing nesting sites all over the county—there are about 23 different nesting sites that are protected nesting areas—as a result of that there are over 4,000 of these terns now that come to San Diego County to nest and make a family.

SR: How does using decoys help promote the increase of the tern's population around San Diego?

Redfern: Decoys have been used in duck hunting for hundreds of years. However, if you're a conservationist, you can use a decoy to attract birds to protected areas. One of the management strategies is to take these decoys and place them at sites like North Fiesta Island. When the birds are migrating north for their breeding season, they look down and see a group of these decoys on the ground. They'll come down and basically colonize [the area] because they nest in colonies. So, it's a way for us to try to direct them to the right spot. Otherwise, they could go and nest other places where there isn't any protection for them. Then they won't be able to breed successfully.

SR: How are these decoys made?
Redfern: Least tern decoys can be made in a variety of ways. The most popular way to make a decoy is a very simple paper maché project that just about any school-age child can do. You start out with some wire—simple coat hanger wire—and you create a cardboard armature that is wrapped around the wire, and then you can use paper maché to fully form the body. The decoys are then painted and set into place.

SR: How long does each decoy last?
Redfern: Each will last several seasons. We don't keep them out all year. We put them in the nesting area where we want the terns to nest and when they're done nesting at the end of the summer, we store the decoys for the wintertime. They're out there a good three or four months.

SR: Who determines the placement of the decoys? And how?
Redfern: Where the decoys are placed is determined by a management group of professional biologists, which include staff from the U.S. Fish and Wildlife Service, the California Department of Fish and Game, and then the folks that are managing the local site, depending upon who the local entity is. It could be a military installation, like the Navy down in the Silver Strand area and North Island; or it could be the City of San Diego. Those folks get together before each nesting season, look at the nesting trends, and then think about other sites that the terns have not been using over the last

couple of years and maybe try to get the birds to come to additional sites that have been provided for them.

SR: What is the hardest part of protecting the least tern?
Redfern: They are not very visible because they're only here during the summer. They're primarily located in protected nesting areas. A lot of these areas are not accessible to the public because they are on federal property. I think the public needs to know more about the species: more visibility and understanding about them; more television crews coming out to welcome the least tern every summer would be great, so everyone knows what a California least tern is.

LEAST TERN

Majestic as you are,
Your crown of black worn proudly atop your head,
Those who show such beauty are rarely found.

Nest upon the sands of the bay,
She will show her hospitality to you,
With arms of nature outstretched and welcoming.

Return, fair bird,
For here you will always be welcome,
And with the changing of seasons,
your home will
be here once again.

— Nathan Wallace

Elegant Tern
(*Thalasseus elegans*)

The graceful form and flight of the elegant tern is only matched by its resilience. This medium-sized tern was first found nesting in the United States in San Diego Bay in 1959. It is primarily found along the Pacific Coast of North America and because of this, it was one of the few North American terns not described or sketched by John James Audubon. Specific nesting locations of the tern are Isla Rasa and Isla Montague in Mexico's Gulf of California, and South Bay Salt Works, Bolsa Chica, and Pier 400, Terminal Island in Southern California. (West and Unitt 2007) Although the elegant tern has been heavily affected by predation and human encroachment, it is still not considered endangered but a "Species of Special Concern" in California. (Collins 2006)

In appearance, the elegant tern is similar to its close relative the royal tern but has a smaller body and a longer, downward curved bill. However, unless a birdwatcher has binoculars or a spotting scope, it might be difficult to differentiate between the two. The bird's call is a clue: the elegant's being more fluid and somewhat slurred and not having distinct notes like that of the royal. ("Terns" 2008)

The elegant tern has an average wingspan of more than 100 centimeters—nearly double that of the California least tern. A patient bird watcher may identify the bird by its crest, which is the longest of all the crested terns, extending from the back of the head to the eye. By contrast, the royal's crest is shorter, ending with all-white feathers around the eye. The elegant's beak appears to droop slightly at the tip and has a pinkish underside, which is another distinguishing characteristic from other terns that have a redder tint on the beak's underside. It is also useful in differentiating it from the Caspian tern with its reddish, straighter and stouter bill. During much of the year, the adult elegant has a white head with a black mask extending from the eye to the nape. In summer though, its appearance changes slightly to a black cap with a long crest. ("Elegant Tern" 2003)

The elegant tern's diet consists entirely of fish. In San Diego this includes the northern anchovy, top smelt, bay anchovy, jack smelt, jack mackerel and long-jawed mudsucker. When foraging, the elegant tern will hover over shallow water before making a swift, graceful headfirst dive into the water, often resulting in dinner in its beak. (Collins 2006)

Historically, the majority of the world population breeds on Isla Rasa, a relatively small island in the Mexican Sea of Cortez. (Unitt 2004) In Southern California it breeds primarily at Bolsa Chica (near Long

Beach) and at San Diego's South Bay Salt Works; many elegant terns were observed there during the course of this study.

Breeding season for the elegant tern generally begins mid-April and ends in September. The bird courts and forms pairs while on migration and away from the nesting colony. When the pair arrives at the nesting grounds, they continue courtship until they are ready to nest. In San Diego Bay the elegant generally nests in clusters adjacent to the Caspian tern. Like many other tern species, it creates a shallow scrape on the ground for its nest. The typical clutch size is one egg, which is incubated by both parents. After hatching, the young aggregate into communal groups.

Fledglings will follow their parents far from the nesting colony in order to be fed and can be observed crouching low and squawking in front of one of their parents, begging for handouts. (Collins 2006)

After breeding season, the elegant tern is found farther north—northern Washington and British Columbia. (Baughman 2003) (Unitt 2004) ("Elegant Tern" 2007) Because the elegant tern lives primarily along the Pacific Coast, it must deal with climate changes such as El Niño activity. After the horrific El Niño event of 1998, there was some impact on the survival rate of elegant tern chicks. Now, almost ten years later, the elegant tern population is thriving. (Collins 2006)

DECLINE OF THE
Elegant Tern

The elegant tern faces many of the threats common to colonial nesting seabirds. However, these threats are particularly detrimental to this species because it breeds in only five nesting colonies; a threat to any of these is easily magnified and reflected upon the entire breeding population. Degradation and disturbance of its nesting sites have previously damaged this highly vulnerable species and have put further restriction on its already limited breeding. ("Elegant Tern" 2003)

Human disturbance is one of the leading causes for the decline of the bird. At the species' main colony on Isla Rasa in the Sea of Cortez, this tern was the target of egg harvesting. Local fishermen had practiced this destructive foraging for many years. First, all eggs, chicks and nests created during the first wave of breeding would be destroyed; within a couple weeks, a second wave of breeding would take place and new eggs were laid. The fishermen would then collect these eggs and sell them in local markets. During the 1950s and 1960s, egg harvesting became more popular than ever, solidifying itself as an important part of the Mexican market because people believed that the eggs possessed both medicinal and aphrodisiac properties. Consequently, elegant tern hatching rates became dangerously low. Though the extent of egg harvesting has drastically decreased over the years, the practice still threatens the Isla Rasa population. (Boswall and Barrett 1978) ("Elegant Tern" 2007) (Ezcurra 2007)

Other threats the elegant tern faces on Isla Rasa is the constant disturbance by tourists of its breeding areas and the extensive mining of guano. Every year, hundreds, if not thousands, of tourists flock to this little island to witness the elegant tern and other bird species breed and raise their young. Often, visitors will walk through the colonies for up-close encounters with the majestic bird. However, as it nests on the ground, simply by walking into a nesting site, visitors are likely to step on some fragile eggs or nests if they are not careful. In 1978, a group of scientific researchers observed a dozen tourists destroy more than 30 eggs within minutes of entering a nesting area. Despite other efforts to protect the bird, agencies have failed to do the obvious and create safe, roped-off walking trails for tourists, or even restrict access to nesting sites during breeding season. (Boswall and Barrett 1978) ("Elegant Tern" 2007)

In other parts of the world, land development has destroyed vital habitats where the elegant tern has been known to make its home. Historically, extensive land use throughout Southern California has left very few breeding areas for the species. Residences, beach-front hotels and other properties have taken priority over the conservation of essential habitats. This damage has often led to the displacement of the tern's population, forcing it to adjust in new, urbanized environments. Not only does the bird have to adjust to a new environment, but it also faces new threats that come with life in an urban landscape. ("Elegant Tern" 2007)

One example of an adjustment the elegant tern has made is nesting on barges off the coast of Southern California. Sadly, this nesting behavior led to the destruction of hundreds of chicks. In June 2006, employees of Point Loma Maritime Services of San Diego were accused of animal cruelty for causing hundreds of tern chicks to flee their nests and drown. On two anchored barges in Los Angeles and Long Beach, dozens of elegant and Caspian terns had nested. However, the constant stress of humans boarding the barges and moving them around the harbor caused the immature tern chicks to panic, flee overboard and drown. The California Department of Fish and Game reported that over a three-day period, 400 to 500 tern chicks perished. Rebecca Dmytryk, a wildlife rescue specialist involved in the recovery of the animals commented, "In all of my experience of seeing injured and abused wildlife, this was the worst." Though many elegant terns have adapted to relocation, others continue to face the annual challenge of finding safe, suitable breeding areas.

Today, many predators, many of which are introduced species, have been known to eat the elegant tern and its eggs, plunging the species into further decline. Because of predator habitat loss, these animals are being pushed into the areas where the tern lives and nests. These unwelcome visitors include feral cats and dogs that prevent the elegant tern from successfully reproducing. The knowledge of a looming predator can stress the tern and cause it to delay egg hatching or even abandon the nest. In the long run this will have a severe effect on the population—perhaps to the point of extinction. Because the elegant tern nests on the ground, its eggs are easily accessible to any predator. If eggs are destroyed early in the breeding season, some terns will lay one to two more eggs. ("Elegant Tern" 2007) (Patton 2007)

Other threats, like the loss of natural fish stocks, have been extremely detrimental to the elegant tern. The Pacific sardine (*Sardinops sagax*) had been overfished, resulting in a food shortage for the elegant tern. However, this was not the only reason for the birds' food shortage. Dr. Enriqueta Velarde connected the loss of natural fish stocks to El Niños that, over the years, have caused the ocean's surface temperature to rise. As the surface temperature gets warmer, the Pacific sardine and other fish swim deeper for cooler water so the tern is no longer able to catch them. When the ocean becomes warmer than average, the tern cannot feed its chicks and may starve to death, causing a further decline in the tern population. (Velarde, et al. 2004) (Ezcurra 2007)

THE ELEGANT TERN

As I sit to ponder
the meaning of wonder,
A timid small man
comes into vision.
Stealthy movements, perfect precision.
Fishing or searching,
I can't tell.
His little black knit cap
fits snug and well.
Delicate and powerful,
graceful and energetic,
he glides across the water,

Nothing could bother
this content soul.
Swooping and sweeping,
scouring for a glimmer,
he feasts on fish.
His quaint song of joy
like a melody to his life.
But what does he say?
If his calls were words
Could we interpret his perspective?
But alas, I can only sit and wonder.

— Paul Christiansen

RECOVERY OF THE
Elegant Tern

Because of the likes of biologist Enriqueta Velarde and the elegant tern's own ability to relocate to undisturbed nesting and survive large fluctuations in food supply, this beautiful sea bird has come back from near endangerment. In 1959 there were a mere 30 pairs of elegant terns in California. Since then, the elegant tern population has increased to 10,000 pairs at South Bay Salt Works in 2003. That is almost a 300-fold increase in approximately 50 years. This is indicative of the elegant tern's nature as a transitory colonial breeder and may also be a sign of hope for the bird's recovery. Due to the overall population recovery, the International Union for the Conservation of Nature (IUCN) now regards the elegant tern as a species of "Lower Risk, Near Threatened." (Collins 2006)

In the past, there was a dire need to help the elegant tern because of its selective and scarce breeding grounds (i.e., the five global locations mentioned in previous sections). Since the elegant tern prefers to be in extremely large groups, the majority of its population breeds on Isla Rasa. If any disruption were to occur on the island, it could result in a serious problem for the species. In the 1970s, such a problem faced the tern population on Isla Rasa. Harvesting of the sea bird's eggs was a local business that eventually became an important staple in the Mexican market. Fortunately, Enriqueta Velarde, then aged 22, was able to decrease this practice. Now, almost 30 years later, she continues to research the elegant tern. (Toropova 2007) (Ezcurra 2007)

Velarde's work began as a simple research project to survey and analyze the tern's population on Isla Rasa, but later expanded to understanding threats that were affecting the bird's population. She became alarmed by the environmental decline of the Sea of Cortez as a habitat for a multitude of creatures. Velarde noticed that sardines were being drastically overfished, thus forcing the elegant tern to eat anchovies, a less nutritious food. In 2003, she released a report that discussed this issue, as well as a natural phenomenon that also affected the tern—a correlation between its sporadic population collapse and the sea's surface temperature. During El Niño years, she discovered that newly hatched elegant terns nearly starved to death because their parents were unable to catch any fish. This was because during El Niño events, sea-surface temperatures rose and the fish would swim deeper for cooler water. (Toropova 2007) (Ezcurra 2007)

As to local recovery of the elegant tern, South Bay Salt Works has played a key role as it is one of the main destinations for the elegant tern during its annual migration. In 2004, there were more than 10,000 elegant tern pairs nesting there. This number may seem extremely high, especially when compared to 1994 when there were only 160 elegant tern pairs at the Salt Works, but the site contains numerous small dikes that help to keep some predators at bay. The U.S. Fish and Wildlife Service's predatory control program played a role in this effort by relocating a pair of nesting peregrine falcons to an adjacent levee. (B. Collins, personal communication 2007)

The explanation behind the large elegant tern population increase between 1994 and 2004 can be attributed to the expansion of its nesting habitat in Southern California to include Terminal Island, a 15-acre site in the Los Angeles Harbor. Unfortunately, the arrival of the elegant tern became a problem because the bird began to displace the endangered California least tern. To solve this crisis, a company named Mad River Decoy created a collection of decoys to encourage the elegant tern to relocate. The project had a degree of success when the elegant tern moved from Terminal Island back to South Bay Salt Works in San Diego. However, while several thousand terns nested there, many abandoned Salt Works' ponds in 2007 because of predator disturbance. (B. Collins, personal communication 2007)

Further research into the bird's basic biology, new regulations in Mexico, and novel recovery efforts will hopefully continue to support this, the most elegant of terns.

Gull-Billed Tern

(Gelochelidon nilotica vanrossemi)

The gull-billed tern has only been a resident of San Diego since 1985. The first few nesting pairs arrived at the Salt Works in San Diego Bay in 1987. Although the population has grown since then, in 2007–2008 only 53 pairs were known to be nesting at the Salt Works. The only other known nesting site for the gull-billed tern in Southern California is the Salton Sea, home to approximately 155 pairs. (B. Collins, personal communication 2008) The population is by no means thriving, and therefore a watchful eye is needed to keep the bird from being on the Endangered Species List. (Unitt 2004)

The gull-billed tern is an average sized tern but its behavior is far from average. It measures about 34 centimeters from head to tail, with a wingspan of 91 centimeters. When it is fully matured, the gull-billed tern has the whitest body of all the terns, with a pale gray back and upper wings and a black crown. Its short black beak is distinctive and aids in identification. This is useful for capturing insects in flight. In the winter, the black crown on its head is replaced by a small black streak around the eyes, and juveniles look much like adults during this time of year.

From the beginning of March through mid-September in San Diego Bay, the gull-billed tern makes its habitat among the sandy shores near South Bay Salt Works.

The bird lays its eggs on bare, open sand. However, unlike many other tern species, the gull-billed tern marks its nest with small bits of vegetation, pebbles, and other natural materials. Eggs are laid between mid-April to late July, with the majority of fledging occurring in July. (Robbins, Bruun, and Zim 1966) (Unitt 2004)

This bird is rarely seen diving in and out of the water in search of fish like so many of its relatives. Instead the tern has a broad diet that includes small insects, lizards and marine invertebrates. During the course of this study the gull-billed tern was observed making its graceful and rapid, steep dive over the sand along the river mouth of the Tijuana Estuary. As the water from breaking waves would recede along the shore, the tern would dive to pull mole crabs from the sand.

A gull-billed tern with a large mole crab flies over the Silver Strand.

However it also possesses an atypical predatory characteristic that is unusual to most birds. One source of its food is the chicks of its relative, the least tern, contributing to declines in least tern numbers. This has created a difficult situation for conservationists; the source of one animal's endangerment is another animal, which could also become endangered. A gull-billed tern was seen preying on a least tern chick and a Western snowy plover chick at the Tijuana Estuary river mouth. In 2004 the California Department of Fish and Game (CDFG) reported that gull-billed terns accounted for nearly 8% percent of the predation on least tern eggs, chicks, fledglings and adults. Since isolating the two species from each other could prove to be an almost impossible task, scientists and environmentalists may be faced with the dilemma of either saving the gull-billed tern or the least tern. ("Gull Billed Tern: *Sterna nilotica*" 2007) (Molina and Marschalek 2003)

Helpless least tern flying above, attempting to save its young.

Decline of the
Gull-Billed Tern

The North American gull-billed tern has become alarmingly small in numbers, plunging the already minute San Diego population into grave danger. Most of the gull-billed tern population is located primarily on the Eastern and Southeastern Coasts of the United States, with only a small fraction located in two different areas in Southern California and Mexico, all of which belong to the Western subspecies *vanrossemi*. However, because it is a cosmopolitan species, the primary reason for the gull-billed tern's dramatic decline was the U.S. East and Gulf Coasts' millinery trade of the late 1800s and early 1900s; the industry was fueled by a demand for the tern's beautiful white plumage. (Molina and Erwin 2006)

Since then, the population has never fully recovered. A problem with determining trends in its population is that, despite a general stability in the bird's overall numbers, locally these tend to be erratic. ("Gull Billed Tern: *Sterna nilotica*" 2007)

Despite its partial recovery since the millinery trade, today the gull-billed tern is severely limited by the lack of suitable, undisturbed habitat—a direct result of human development. The tern tends to be susceptible to disturbance, and it is even known to abandon its nests when there is trouble. For example, some former nesting areas in salt marshes have been abandoned because of human encroachment. Human development also causes juveniles to leave the nest before they are ready, exposing them to harsh environmental conditions and predators. The gull-billed tern is also susceptible to pollution and pesticides that result in problems such as eggshell thinning and reproductive failure. ("Gull Billed Tern: *Sterna nilotica*" 2007) Furthermore, since the 1970s, gulls have begun to take over gull-billed tern colonies and nesting grounds. Other threats to the population include the scarcity of winter food, flooding and predation. The gull-billed tern, like many other species, faces a long recovery process that needs to be tackled from many angles.

The gull-billed tern is one of the only terns to feed on infaunal invertebrates while diving along the shore.

RECOVERY OF THE
Gull-Billed Tern

Despite human development and other environmental concerns, the gull-billed tern is currently not considered endangered by state or federal agencies. However, it is on the CDFG Bird Species of Special Concern List. This means that it is a fragile species, and measures must be taken to conserve the population. If any further harm is done to this bird, it could become endangered. Ultimately, protected areas may be the only possible solution to saving the birds. (Unitt 2004). Presently, the National Wildlife Refuges in San Diego Bay and Salton Sea protect nesting gull-billed tern colonies. The sole nesting area in San Diego for the gull-billed tern is on the South Bay Unit of the San Diego Bay National Wildlife Refuge where it receives the same protection, management and monitoring efforts as the least tern.

In the past, the only measure taken to save the gull-billed tern was a series of laws that made the millinery feather trade illegal. The first of these laws was the Lacy Act of 1900, which prevented the interstate trading of any animal that was protected under the state, fining anyone who did not abide by the law up to $1,000. Unfortunately, people found ways to work around it, and the birds continued to be poached and killed. The end of the millinery feather trade and the deaths of thousands of gull-billed terns can be directly credited to the Migratory Bird Treaty Act of 1918. This law closed these loopholes by protecting all native migratory birds. Even so, the birds are still barely recovering from these events. (U.S. Fish and Wildlife Service 1918)

The future of the Salton Sea restoration project could be vital for the gull-billed terns of California. If this primary nesting habitat continues its downward spiral because of water quality and avian health problems, the role of the Salt Works nesting area, and the creation of other protected nesting areas for the bird, will become extremely important.

CHAT WITH A GULL-BILLED TERN

As I strolled along a marsh,
A cry of help I heard
It came from a little gray and white bird.
"Oh, little gull-billed tern,
You really need to learn
The world is hard
And you don't come first.
We have wars to fight,
Cars to make,
Police sirens always can be heard.
We don't have time for you, you stupid little bird!"
Disheartened, the bird replied,
"I shouldn't be denied!
Just listen here,
I'll make it clear
You may not understand.

Even a little bird like me
Is important for biodiversity
Like every creature in the land!
I eat others,
Others eat me
It's an important role in society!
Protecting me is essential.
I truly need to stay.
And if you try
So sure am I
That even you can be a steward of the bay!"
And with that,
The little bird flew away.
But as I left the marsh,
My feelings strained,
Its little gull-billed words remained.

— Sara Islas

103

Light-Footed Clapper Rail
(*Rallus longirostris levipes*)

The light-footed clapper rail is one of the rarest birds along the West Coast of North America. In recent years, it has become a mascot for both wetland and endangered species conservation. The rail is native to the coastal salt marshes spanning from Southern California to southernmost California and the Pacific coast of the Baja peninsula in Mexico. It is considered to be an "indicator species," which means that a healthy clapper rail population is usually indicative of a healthy wetland community. This has become a useful gauge for conservation biologists and policymakers alike. In recent years, the rail has become one of the most endangered birds in California, due primarily to habitat loss. (Unitt 2004)

Although the clapper rail prefers solitude, it is easily distinguishable when it emerges from hiding by its long, orange beak, red-orange breast, spotted or striped brown body and its long legs and toes. Clapper rails are opportunistic omnivores and will eat small fish, insects, certain plants, seeds and other available food. The rail, which is about the size of a small chicken, uses its curved beak to probe mud and water in search of small crustaceans such as striped shore crab and mollusks like the olive ear snail. Though it prefers to travel on foot, the rail is capable of flying, swimming and diving when necessary. The rail's indigenous predators include raptors, herons, coyotes and foxes. However, the worst threat to the clapper rails comes from introduced species. (Zedler 1992)

The rail has an interesting relationship with Pacific cordgrass (*Spartina foliosa*), a saltwater marsh plant. It builds canopy nests from cordgrass that serve as shelter and protection. The ingenious design is engineered to rise and fall with the tide, allowing the nest to remain safe and accessible at all times. With its lean, streamlined body, the rail is able to move quickly through thick areas of cordgrass to seek protection from danger. John James Audubon described the bird's ability to navigate dense marshes: "Clapper Rails have a power of compressing their body to such a degree as frequently to force a passage between two stems so close that one could hardly believe it possible for them to squeeze through." (Audubon 1842)

The clapper rail breeding season occurs from January through April and is accompanied by a period of vocalization and courtship. This breeding period is an ideal time to make rail population estimates because of the distinctive calls the bird makes during courtship. The male's call is described as a *kek kek kek kek* sound, quick and disjointed, similar to the clucking of a chicken. The female responds with a *kek kek kek kek burr* sound, the interval between her clucks decreasing to become a solid fading note.

Female rails lay four to 10 eggs per clutch. Occasionally, pairs may have more than one clutch during each mating season. The male and female typically take turns incubating the eggs and collecting food. The eggs then hatch after 21–24 days of incubation. Nestlings become independent by the age of six weeks and reach sexual maturity in approximately one year. Clapper rails usually survive no longer than four years in the wild, but have been known to live more than 10 years in captivity. (Unitt 2004) (Vanner 2002)

Decline of the
Light-Footed Clapper Rail

In 1973, the light-footed clapper rail was officially recognized as an endangered species. The primary cause for its dramatic drop in population is the destruction of California's coastal saltwater marshes. It has been estimated that more than 90% of the saltwater marshes that once dominated coastal California no longer exist. Under such circumstances, it is not surprising that the rail population experienced a swift decline over the past few decades. Additionally, local habitat loss is particularly damaging to the species because as a nonmigratory bird, it does not disperse when its habitat is compromised. ("About Our Cause" 2007) (Unitt 2004)

While habitat loss is the primary cause of the rail's problems in the past decades, other factors have contributed as well. Introduced species such as red foxes have had a devastating impact upon the rail. Cats and dogs (both feral and domestic) are serious threats to it as well. The small bird has no defenses against these unnatural predators and perishes quickly when such animals invade. Clapper rail parents, unable to defend their eggs or nestlings from large predators, will abandon their nests when threatened, leaving their young with little hope for survival. While rails can produce a second clutch to replace failed nests, this requires great time and energy and is not possible in the latter part of the spring. (Gailband 2007)

Another compounding problem for the rail is that while its population decreases, populations of its natural and unnatural predators have undergone little change. As a result, for example, if a bird-of-prey

killed a clapper rail it would not be a problem under normal circumstances. However, a natural occurrence such as this is a great loss when the bird's entire population is composed of only a few hundred breeding pairs. Additionally, the health of Pacific cordgrass, the plant species upon which the rail is largely dependant, is declining in California. This decrease, also caused by human development, eliminates an important source of protection, nest-building material and even food for the bird. (Zedler 1992)

With human development along most of California's previously marshy coastlines, these areas have not only decreased in size, but become fragmented. Instead of long, almost continuous stretches of wetlands along the coastline, California now has only small, isolated marshes. This prevents rails from having as many breeding opportunities, creating what is called a "genetic bottleneck." This phenomenon of low genetic diversity is one of the bird's most serious current threats, as inbreeding could ultimately prevent the establishment of a stable clapper rail population. (Gailband 2007)

ODE TO A CLAPPER RAIL

Light-footed clapper rail
The puzzled observer ponders you
With your peculiar habits
Your confused, chicken-like gait
Your discordant, cacophonous call
Your frantic, unpracticed flapping
Hiding in your gasping, choking, wheezing, drowning swamp
With your helpless struggle to survive for even one more day.

Light-footed clapper rail
I have seen you
With your charming character
Your confident stride
Your song echoing to the heavens
Your determined flight
Smiling in your steadily breathing marsh
With your courageous conviction to celebrate the next dawn.

— Sean Curtice

RECOVERY OF THE
Light-Footed Clapper Rail

While there are many forces negatively impacting the light-footed clapper rail, there are fortunately several recovery plans in place for the endangered bird. In late 1998, the Chula Vista Nature Center (CVNC) located on the shores of San Diego Bay began the Light-Footed Clapper Rail Recovery Program. The project was designed to breed and release rails into the wild to increase their overall numbers. However, after two breeding seasons (1999 and 2000), the single pair residing at the center had yet to breed. This led to a reevaluation of the program's methods. Those involved with the program began to wonder whether captivity may be too stressful for the birds, if there were possible dietary issues or if the pair simply were not compatible. (Gailband 2007)

In 2001, a second pair of male and female wild rails were added to the breeding program. The original couple was split up, creating two new pairs. It was hypothesized that having two pairs on site would fuel the rails' territorial tendencies, producing the necessary hormones to encourage breeding. The predictions proved to be correct with the first successful rail hatchings that year. This success also marked the beginning of SeaWorld's involvement with the program, when healthy chicks were transferred to their facilities to be hand-reared before being released. That year, seven clapper rails were released into the wild and two were kept at the CVNC to form a third breeding pair. (Gailband 2007)

An identification band is attached to a clapper rail before it is released into the wild.

Since this first successful release, the CVNC and SeaWorld have continued the clapper rail population rehabilitation program with great success. By 2007, more than 171 birds had been released in nine significant coastal marshes. Additionally, the total estimated population of wild rails has risen from 201 breeding pairs in 2001 to a record high of 408 breeding pairs in 2006. With working protocols established by the CVNC and SeaWorld, the San Diego Wild Animal Park has also joined the breeding efforts. As of this writing, each of the three organizations has two breeding pairs. ("About Our Cause" 2007) ((Mannes 2007)

All involved with the rail project were optimistic about the findings of recent population estimates, but it wasn't solely the efforts of the project that have helped the bird's population increase. In the words of Dr. Richard Zembal, the foremost expert on the light-footed clapper rail and leader of the Clapper Rail Study Team, "Conservation efforts, such as salt marsh restoration … benefit the Clapper Rail and its environs [to] enhance our rich coastal wetlands …." Thus, preserving natural habitats remains the primary way to help endangered animals survive. (Zembal, Hoffmann, and Konecny 2007)

While the rail population still has a long way to go before its population returns to previous conditions, the future looks bright. With new partners in the project, including the San Diego Wild Animal Park, recovery efforts are stronger and more prevalent than ever before. Additionally, further research is in place to improve radio telemetry processes to help track released rails, thereby affirming the effectiveness of the program. The clapper rail project also hopes to expand to Mexico in the near future, since it is believed that populations in the Baja California region could play an important role in the preservation of the species as a whole. (Gailband 2007)

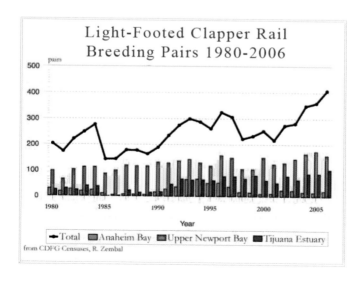

Recent surveys indicate an increase in wild clapper rails. Courtesy Dr. Richard Zembal. ("About Our Cause" Clapper Rail Study Team)

Bird Surveying

It was a cloudy afternoon at the Tijuana River Estuary as I stood with several other volunteers and biologists, getting ready to begin a survey of the local light-footed clapper rail population. For someone who had been studying clapper rails as I had in the past few months, this was an amazing opportunity. Richard Zembal, the foremost expert on the species and Charles Gailband, curator of birds at the Chula Vista Nature Center, were presently explaining the process of counting the birds. "Clapper rails are very secretive," they said, "and are rarely seen in the wild, even by the most avid birdwatchers." Because of this, surveyors had devised a special means of counting the animals that differs from the visual method for assessing the populations of most other bird species. Because clapper rails are territorial, when an individual or a breeding pair of birds sounds their signature "kek kek kek kek" and "kek kek kek kek burr" vocalizations, nearby birds respond with the same call. Surveyors have been able to take advantage of this by playing recordings of the rail's call and listening for replies. It is also possible to differentiate the calls of single males, single females and breeding pairs.

Volunteers were each given binoculars and a map of the estuary and told to mark the approximate location of the bird on the map. A breeding pair was to be designated with a "P;" a lone male with a "K" and a lone female with a "KB," representing their respective calls. I was paired with another first-time surveyor, and we were assigned to patrol the main trails rather than wade through the marsh. Though I initially thought this would mean a smaller chance of seeing a clapper rail, I could not have been more wrong.

We began making our way down a quiet trail, pausing occasionally to listen for both the recorded and actual calls of our elusive subjects. The counting system worked brilliantly; even though the birds' voices were often a few hundred meters away from one another, the sounds carried clearly across the coastal wetlands. Once we heard the recorded call, a chain reaction would occur all around us and the rails' vocalizations would rise up from the grasses, responding to the initial noise and then to one another. When we came to the end of the trail, we noticed that there was a watery, open area, just bordering the cordgrass in the marsh nearby. We decided to stop and observe in hopes of a sighting.

After watching a group of plovers and a passing curlew and making a few more notes of suspected rail locations, we saw something very surprising. A bird emerged from the cordgrass and walked along the bank of the tidal creek towards an area covered by foliage. I grabbed my binoculars for a better look and saw to my amazement that is was a clapper rail! Excited by this rare opportunity, I followed its movements carefully as it went behind a patch of vegetation. My colleague and I watched as it returned, accompanied by a second bird and the two then proceeded to mate, shrieking their characteristic breeding duet throughout. I was still in awe when they ran off a moment later. To witness the breeding process of such a reclusive species was a rarity; I may have been one of only a handful of human individuals to have observed this in the wild.

As time passed and it grew darker, I marked more and more rails on the map, mostly breeding pairs but I did spot a solitary rail through the cordgrass on the side of the trail. When the time came, the other volunteer and I returned to the rendezvous point to meet the rest of the surveyors. In total, I had counted 13 pairs plus three lone males and one lone female. I was extremely pleased with the outcome of my efforts and I was not the only one who was surprised by the enormous amount of rail activity seen that day. Zembal later said that, "The calling frequency at Tijuana Marsh that evening was more intense than most of the seasoned observers had ever experienced." After analyzing the surveyors' notes and compiling the annual light-footed clapper rail population report, Zembal estimated 142 breeding pairs in the estuary—an amount that had not been recorded in years.

Though I have since continued working extensively with clapper rails, conducting hours of observations and attending multiple releases, this survey had probably the most impact for me, as it was the first time I was able to move from clapper rail research in books to direct observation in the field. To be a part of a movement to help an imperiled species, especially one that is proving so successful, is truly an extraordinary experience.

Black Oystercatcher

(*Haematopus bachmani*)

Few would argue when naming the bird that is most adapted to the rocky intertidal shores along the California coast—the black oystercatcher steals the title. The piercing call of the bird often announces its arrival, and its contrasting red-orange bill and black body can often be recognized from a great distance.

The black oyster-catcher is a shorebird that stands approximately 38 centimeters high. Both the male and female have bright red bills and dark bodies.

The juvenile oys-tercatcher also has a dark body; but, in contrast to the adult, has a duller, more or-ange bill. The oystercatcher's unique coloring is one of its de-fining features, making it virtually unmistakable to identify. Because the oystercatcher resides near the shoreline, it can blend fairly well with colorful rocks that are encrusted with seaweed and small crustaceans.

Even though the black oystercatch-er appears to be well-adapted to the rocky intertidal environment, it is still threatened by many factors. It is estimated that as few as 8,800 pairs of oystercatchers remain. It can be found on rocky shorelines of the Pacific Coast and north to Canada.

In this region, it is found on Los Coronados Islands (Baja California, Mexico), but its natural habitat of rocky shoreline is limited in San Diego County. Locally, most sightings are along the cliffs of La Jolla, Point Loma, and the Zuniga Jetty, which extends off the southwest tip of North Island at the mouth of San Diego Bay. (Unitt, 2004). Unlike most birds along the Pacific Coast, the black oystercatcher does not migrate and tends to remain in a certain area. (Andres 1999)

Though the black oystercatcher has an unmistakable call, it can be difficult to locate its nest. It eggs are small, brown, and are often mistaken for rocks. Because of this, humans sometimes unknowingly step on the precious eggs and destroy them. (Andres 1999) In San Diego County, nesting has not yet been documented, but courtship and nest-site selection have been observed, suggesting the possibility of nesting. (Unitt 2004)

A unique characteric of the black oystercatcher is that it picks its mate for life. The female lays two to three eggs at a time, and both the male and female take turns incubating the eggs while the other parent searches for food. The incubation period is four weeks. When the offspring are hatched, they stay with their parents until about five weeks of age when they venture out into the world on their own. At this stage, the maturing chick is dark brown with an orange bill. To make it this far, the chicks have avoided reckless humans, dangerous predators, and the unpredictable weather to become a healthy bird. ("Black Oystercatcher *Haematopus bachmani*" 2007)

Since the population of the black oystercatcher is so widespread, each colony has its own threat. Some are more prone to human interference while others may be bothered by egg-consuming predators such as river otters, foxes and eagles. Generally the bird has an average life expectancy of about 15 years. (Andres 1999)

Despite its name, the black oystercatcher primarily feeds on mussels, limpets and other marine organisms; its bill is shaped conveniently so it can pry open shells for food. It is sedentary in nature, so food must be close to its nest. However, it is rare to see a black oystercatcher feed because it is sensitive to any human activity.

The oystercatcher is a keystone species whose numbers are declining. Therefore, it is vital that we conserve and restore this amazing bird for the health of our environment. (Tessler, et al. 2007)

DECLINE OF THE
Black Oystercatcher

The U.S. Shorebird Conservation Plan lists the black oystercatcher as a "Species of High Concern," due to its relatively low numbers. One of the reasons for the oystercatcher's population decline is disruptive human activities, which hinder the bird from breeding, and frequently cause it to abandon nesting sites. This leaves the eggs open to predation by sea gulls and the effects of the blistering sun.

An additional cause of its population decline is predation by animals such as foxes and rats. In wetland areas, there were once islands that allowed the bird to breed safely without fear of predators. However, human activities combined with erosion have caused many of these islands to disappear allowing predators easier access to oystercatcher nests.

The black oystercatcher's small population makes it at risk to large-scale disturbances, such as oil spills. The 1989 *Exxon Valdez* oil spill in Alaska's Prince William Sound had a detrimental effect on breeding oystercatchers in that area, as well as along the entire coastline. Because of the spill, 20% of the bird's population in the area was killed. The breeding activity along the heavily oiled coast was disrupted by 39% percent, and

chick survival was lowered as well. Oil spills in general affect the black oystercatcher's overall well-being and more so when they occur in the species' habitat. ("Black Oystercatcher (*Haematopus bachmani*)" 2007)

RECOVERY OF THE
Black Oystercatcher

The black oystercatcher is a fragile species. However, there are few focused efforts in place to restore its population. Direct conservation efforts for the oystercatcher are limited by the lack of baseline population numbers. Addressing such research gaps as fledgling success, local population status and trends will increase the understanding of the bird, as well as help develop conservation strategies. Some scientists are tagging the bird to monitor how much its population is fluctuating.

There are techniques available to help the dwindling population. Researchers have found that placing black oystercatcher decoys on potential nesting grounds attract birds to rocky shorelines.

Monitoring of the bird's nests has also been started in places like Alaska's Kenai Fjords National Park. From these studies, scientists hope to interpret different threats to the black oystercatcher's reproduction and rearing cycles.

Education of the public is yet another key factor. Two groups in California that are good at educating the public with their captive bird programs include the Chula Vista Nature Center and the Monterey Bay Aquarium. When visiting breeding grounds, the public should be cautious around eggs so they are not accidentally disturbed or destroyed. It is also important for people to avoid bringing their pets into these areas because that presents a further threat to the oystercatcher. Through the continued efforts of people and support of concerned organizations, recovery is possible for this unique species.

Western Snowy Plover
(*Charadrius alexandrinus nivosus*)

The Western snowy plover is one of the most threatened shorebirds in the United States. This petite and rare plover finds no better home than San Diego's famous sandy coastline, which is unfortunately a perfect place for human interference and habitat destruction. The Western snowy plover primarily nests in pairs in coastal regions. It ranges from 15 to 17 centimeters tall and typically weighs 34 to 58 grams. Its coloration consists of a white underbelly, a tan top and black markings on the neck, head and above the beak, allowing it to blend in with its surroundings. The plover lives along the California and Mexican coastlines, but has been sighted as far north as Washington. San Diego currently has the largest concentration of the species in the United States, with 207 plover sightings during the Audubon Society's Christmas Bird Count in 2006. Its diet consists mostly of insects such as brine flies—small flies that are commonly found on beaches around mounds of kelp. However, it will also occasionally feed on small crabs or other invertebrates. ("Snowy Plover" 2007) (Unitt 2004)

Western snowy plover nesting grounds are typically located on sandy beaches and dried mud flats. Like all plovers, it places its eggs on the bare sand; the eggs are speckled with brown and tan, so they can be easily camouflaged in the sand and normally range from one to three eggs per

nest. During mating season, the female plover lays her eggs and then leaves the nest-guarding responsibilities to the male. When a nest is approached by a predator, the male bird will act as though it has a broken wing and will then attempt to lead the predator away from the nest. If the predator is not deterred, the plover will flee the nest. This technique has worked in deterring predators for thousands of years, but unfortunately it is not helping the plover to survive human disruptions, and the bird is currently facing grave declines in population. (U.S. Fish and Wildlife Service 2006)

DECLINE OF THE
Western Snowy Plover

Along with the California least tern, the Western snowy plover has become the poster bird for threatened coastal dunes and sandy beach habitats in California. The portion of the species that nests on coastal beaches and adjacent bays was protected under the Endangered Species Act in 1993. There are many causes for the bird's decline, but the main issue is habitat alteration occurring through development. Every year more and more habitat is being lost to the building of beach-front condos, houses, restaurants, hotels and paved parking lots.

The reproduction rate of the plover has also been affected. Not only have human disturbances, predation and habitat loss affected its nests and current incubations, but the plover has been discouraged from nesting altogether. Another reason is the increased use of beaches in California, via off-road vehicles and beachcombing. The U.S. Fish and Wildlife Service has observed that the nesting season for the plover (March to September) coincides with the most active recreational beach activity. These activities can disturb or even destroy plover nests, but not all of the causes of plover population decline can be directly attributed to humans. Invasive species such as foreign grasses have reduced the amount of available breeding grounds for plovers; and predation by animals such as dogs, cats and red foxes is responsible for the destruction of many nests. ("The Snowy Plover Page" 2007) (U.S. Fish and Wildlife Service 2007)

As for population size, between the years of 1977 to 1980 there were approximately 1,590 Western snowy plovers in California. For the period 1989–1991 the population had dropped to about 1,370 individuals and by 2000 it had further fallen to an alarmingly low number of less than 1,000. After this disastrous decline, the California population thankfully began to increase—in 2002 there were 1,387 recorded observations, followed by 1,444 in 2003, 1,904 in 2004 and 1,680 in 2005. In 1970 plovers nested at more than 50 sites in the state, but today there are fewer than 20 of these sites left. If even one of these nesting sites were to be lost to development, it would be potentially disastrous for the Western snowy plover population in California. (Tijuana River National Estuarine Research Reserve 2007) (U.S. Fish and Wildlife Service 2007)

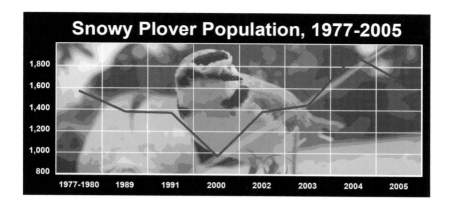

Snowy Plover Population, 1977-2005

| | 1977-1980 | 1989 | 1991 | 2000 | 2002 | 2003 | 2004 | 2005 |

Day-old snowy plover hatchlings sit on a nest of shells.

RECOVERY OF THE
Western Snowy Plover

Although this shorebird is threatened, strong activism has kept it from extinction. There are many organizations that have devoted themselves to the recovery of the plover and its fragile habitat. Unfortunately this diminutive plover resides in some of the most sought-after real estate in North America.

From 2000 to 2004, the plover population increased by almost 1,000 from previous numbers, however this declined again in 2005. The majority of the plover population resides in California with nearly 1,700 individuals. Only about 100 individuals were observed in Oregon and Washington combined. (U.S. Fish and Wildlife Service 2007)

This increase is partly due to the establishment of private plover breeding grounds. Many development projects have been halted because they would have been built on plover breeding habitat. The military has also stepped in to assist in the recovery of this shorebird. Plots of land owned by the military are being used for recovery efforts, providing breeding grounds for the plover and other threatened birds. The government has also placed a $2,500 fine for the disruption of a plover nest. Other efforts have included the removal of European beach grass, an invasive plant species that has been shown to lead to a decline in plover nesting activity. ("The Snowy Plover Page" 2007)

Additionally, the U.S. Fish and Wildlife Service, California/Nevada Operations office, has created a Recovery Plan for the Pacific Coast Population of the Western Snowy Plover. The objective of this plan, outlined in September 2007, is to remove the plover from threatened

A snowy plover parent feigns injury to draw a suspected predator away from the nest.

species status by increasing its numbers, ensuring its sustainability and monitoring its population. According to the parameters set out by this plan, there is a possibility the plover may be removed from the list by 2047. This long-term recovery plan truly showcases the need for better care of our environment and the species that reside in it. (Hornaday, Pisani, and Warne 2007)

There is still a long way to go before the Western snowy plover can be removed from the threatened species status. Unfortunately, there has been growing outrage from the public because many people are having to give up their beach activities for the plover. Some angry citizens have suggested throwing "Snowy Plover Stomp" barbecues in protest, an example of some of the more radical movements against habitat restoration and species conservation.

Black Skimmer
(Rynochops niger)

With its unique feeding habits, remarkably vivid red beak and jet black feathers, the black skimmer is truly an incredible bird. When it hatches, the skimmer's top and bottom mandibles are around the same length; but as the bird matures, the bottom bill grows two to three centimeters longer than the top. This distinctive characteristic provides the bird with an especially useful tool as it skims across the water seeking food.

The black skimmer belongs to the *Rhynchopidae* family of skimmers. The bird is usually between 40 and 50 centimeters in length and has a wingspan of 112 centimeters. The female weighs about 200 grams and the male 300 grams. The skimmer's body is a spectrum of bold colors, including an all-black head, piercing black eyes, white neck and brilliant red-orange feet. ("Black Skimmer *Rynochops niger*" 2007)

Unlike most birds, the black skimmer relies more on its sense of touch rather than its sense of sight when hunting for food. This allows it to hunt for small fish, krill and shrimp, during both day and night. During its search, the black skimmer glides its lower mandible across the water; when it feels something enter its mouth, it quickly snaps its beak shut and swallows the prey whole. A parent bird will carry its catch back to its hungry chicks. ("All About Birds: Black Skimmer" 2007)

First-hand observations conducted by students at High Tech High have revealed the majestic beauty of this unique species. "We ventured out to Shelter Island near San Diego Bay to observe the majestic feeding motions of the black skimmer. ...one flew out of nowhere and swooped down across the water. It skimmed across the surface for a few seconds, leaving ripples behind, before it came back up. It repeated this act several times. We sat there in awe as the skimmer sped its bright beak so smoothly across the top of the water."

The black skimmer tends to live in habitats near oceans and bays such as estuaries, wetlands, lagoons and creeks. In California, it has been sighted in Bolsa Chica, Los Angeles Harbor and San Francisco Bay. In San Diego, its primary winter roosting site is located at Crown Point on the shore of Mission Bay. Its primary nesting site is South Bay Salt Works (SBSW), with recent expansion to Batiquitos Lagoon in San Diego County and to the mouth of the Santa Margarita River at the U.S. Marine Corps Base, Camp Pendleton.

Black skimmer eggs; occasionally parents will place odd objects in nests including golf balls and champagne corks.

DECLINE OF THE
Black Skimmer

The black skimmer's partiality towards the coastal climate of Southern California makes the species especially vulnerable to human activity. The skimmer often nests on coastal beaches or inlets, a common place for human activity, so visitors may disrupt the skimmer's habitat by unintentionally stepping on its hard-to-see eggs. In addition, even with DDT being banned for decades, the bird still experiences eggshell thinning and chick mortality. These are caused by sediment contaminants that have accumulated in the fish it consumes not only in this area, but in its Mexican wintering grounds. Further, changes in noise level, such as traffic, can disrupt natural breeding habits of the species. According to the U.S. Fish and Wildlife Service, because of human disturbance "[the black skimmer's] incubation period can be as short as [21 days] but can last [up to 25 days] in areas of disturbance." (Collins 2007)

Research conducted by Dr. Joanna Burger verifies these findings. Burger's study subjected six black skimmer subcolonies to daily and weekly checks by humans. Some birds deserted their original nesting sites and settled in other colonies. As a result, many chicks were abandoned along with the nests. (Safina and Burger 1983)

With Southern California coastal property among the most valuable in the country, nesting habitats free of humans and predators, including the domestic dog and cat, are rare. Because of this, protected areas such as SBSW and Batiquitos Lagoon have been established. Areas that are set aside and free from human interference, such as off-road vehicles, are vital to the survival and conservation of the black skimmer.

RECOVERY OF THE
Black Skimmer

It is estimated that there are currently between 120,000 and 210,000 black skimmers in the world. A global population threshold, a tool used to evaluate population numbers of a species in relation to the rate of its decline, has not been conducted for the black skimmer. Despite this, many scientists believe that the bird is fast approaching its threshold, which is declared when a population declines more than 30% during the course of a decade. Regardless of human disturbance and habitat destruction, there is hope for the black skimmer. As of this writing, the skimmer is a species of "Least Concern" on the International Union for Conservation of Nature and Natural Resources listing. ("Black Skimmer *Rynochops Niger*" 2007)

One major way to stabilize the bird's population would be to make more "human free" land available to the species. Today in Southern California, there are not many places where the black skimmer can thrive. San Diego County's Batiquitos Lagoon is one of few places it can flourish because of the location and relatively disturbance-free zone. In the lagoon, black skimmers mainly breed in the East Basin. In 1995 and 1996, ten pairs nested in the West Basin and then moved to an East Basin island where they nested from 1997 to 1998 and again from 2001 to 2003. In recent years, pair numbers have increased from approximately eight or ten to around 26. (Unitt 2004)

Another habitat suitable for the black skimmer is SBSW, as it provides an isolated habitat and contributes to the skimmer's breeding success. Most of the birds establish their nests on levees that are free of

One-day-old black skimmer hatchling at South Bay Salt Works.

any kind of disturbance. The skimmer's population at SBSW has been stable during recent years. In 1998, there were at least 280 nesting pairs. The population dropped to about 200 pairs in 1999 and then rose to 280 in 2001. The population has increased further, peaking to 752 nests in 2005. (Unitt, 2004)

Sometimes human intervention can help, rather than harm, the black skimmer; people can assist by reporting nesting sites. A program that would train people to identify the skimmer's camouflaged eggs as they walk along the beach would be of great value in reducing egg destruction. Restricting or limiting the use of off-road vehicles is another way to benefit skimmer populations. These are only two examples of how a conscious human effort can help the recovery of a species.

A PICTURE OF A SKIMMER

Little black skimmer.
Be patient and hold still.
All I want is a picture.
Timid as you may,
This is going to be your lucky day,
Your one shot at fame.

I steady my camera
Focusing on the best of you
1..2..3.. CLICK
And you disappear.

As I check my camera
I see nothing but the grayness of the sky.
Little skimmer, how could you?
A perfect picture ruined
By the quickness of your disappearance.

— Jennifer Zarzoso

Long-Billed Curlew
(Numenius americanus)

When one sees a bird like the long-billed curlew standing in the mud or majestically flying along the coast, it seems to radiate beauty and magnificence. The long-billed curlew, a member of the sandpiper family, is one of the largest shorebirds in North America. It is easily distinguishable by its long curved bill, which is four times the length of its head. The curlew got its name from its unmistakable call: *curleeuu, curleeuu, curleeuu.* Today, it is one of the more common shorebirds in San Diego, though its population has been constantly declining. In San Diego, the bird can often be found at the mouth of the Tijuana River Estuary and in the South Bay. It usually lives among grasslands and along mudflats, but can be seen almost anywhere when migrating during the fall. ("The Long-billed Curlew" 2007)

The curlew has an interesting appearance. Its small head is designed for probing the mud, with a bill that may reach 20 centimeters in length, and curving slightly downwards. Its neck is also long, giving ideal leverage when searching for food. The curlew's tiny feet are rounded to adapt to its mudflat or grassland habitat. Its bland coloring provides camouflage against the background of its environment; but its movement resembles a dancer waltzing around the marsh. (Unitt 2004)

The curlew spends much of the day wading and foraging in salt marshes. It feeds on insects, marine and freshwater invertebrates, mollusks, amphibians and wild fruits. It will also eat small saltwater shellfish and all sorts of crayfish. ("The Long-billed Curlew" 2007)

The curlew is found along most of the West Coast of North America. During the breeding season, it ranges from New Mexico northwards to British Columbia, although it has been known to remain in San Diego's South Bay all year long. In the winter it may migrate to Southern California and farther south.

When the curlew mates, the male makes the advances, attracting the female by his intricate flight patterns. If he does a satisfactory job, the female will lay as many as four eggs between April and June. Both parents take turns incubating the eggs over a period of 20 days. Newly hatched chicks are able to run about after several hours and forsake the nest the following day. The young begin to fly when they are about 35 days old. ("The Long-billed Curlew" 2007)

DECLINE OF THE
Long-Billed Curlew

With a current global population estimated at 20,000 birds and considered "highly imperiled" the long-billed curlew is on the U.S Shorebird Conservation Plan. Factors contributing to its decline include hunting, climate change, disturbance of nesting sites and habitat degradation. The curlew's nests and eggs are protected under the federal Migratory Bird Treaty Act, yet fewer than 10% of curlews nest within protected areas. Without further help, it will only be a matter of time before the long-billed curlew ceases to exist.

One of the first major impacts on the curlew population was trapping and shooting. Although the long-billed curlew is reputed to be less tasty than the Eskimo curlew, it was hunted and sold in marketplaces around the world. In San Diego during the late nineteenth and early twentieth centuries, this hunting took its toll. Elsewhere, the bird was often decoyed and captured in its breeding grounds. Today, it is illegal to hunt the curlew, but experts caution that some level of illegal harvesting continues. (Dugger and Dugger 2008) (Unitt 2004)

The infestation of certain non-native weeds and grasses has created a dramatic impact on the curlew's nesting areas. Although exotic grasses, such as cheatgrass, have been known to increase breeding numbers in the Western United States, other exotics like crested wheatgrass and knapweed can cause severe degradation by creating dense, tall vegetation that replaces the spacious, open grassland conditions preferred by the curlew. (Dugger and Dugger 2008)

The damage to winter territories has also affected the curlew population. Over 90% of California's Central Valley wetlands have been drained, and grassland habitat has been paved over or converted to agricultural purposes. The activity from roadways and off-road vehicles disrupts vital parenting behaviors, and some curlew parents may even abandon their nests. Trampling by grazing livestock also causes destruction of curlew nests. (Dugger and Dugger 2008)

The degradation of the curlew's breeding, foraging and wintering habitats has the greatest and most dramatic effect on its population. Land development for housing and other urbanization in San Diego County are eliminating native grassland and mudflat areas that were once home to the curlew. Climate change has also been found to affect breeding habitats, plus debris (e.g., car bodies, dead livestock, large appliances and yard trash) can decrease the appeal of potential nesting areas. (Dugger and Dugger 2008)

The curlew population is estimated to be decreasing annually by 1.6%. By the next century, it may become completely extinct. If action is taken now to preserve and protect the long-billed curlew, there will still be hope for the species.

Invasive species such as the thistle have displaced many native grasses, which provide nesting habitats for curlews.

RECOVERY OF THE
Long-Billed Curlew

The long-billed curlew's population has dropped dramatically with only 7,000 birds remaining along the southern portion of the Pacific Flyway. There has been an enormous loss of wetlands—the bird's essential breeding habitat—and the remaining wetlands must be preserved to help protect the long-billed curlew from extinction.

Agriculture has also had an impact on the curlew population since the bird cannot live in the dry, trimmed and pesticide-sprayed grasses near agricultural sites. In order to survive, it needs lush, wet grasses, with plenty of water and insects. To provide this, portions of wetlands could be sectioned off and turned into wildlife refuges. Novel ways to sustain the curlew population and still meet agricultural demands are necessary. ("Curlew" 2007)

Native grasslands are an essential curlew habitat. The long-billed curlew needs short grass in nesting areas for predator detection and chick mobility. Draining, liming and fertilizing of land should be minimized and cattle grazing limited to certain areas. Controlled burning of land may create a suitable nesting habitat by stimulating grassland growth.

Cattle grazing from late summer and into the fall can result in nesting and feeding areas for the following spring. Grasslands can be retained or recreated by avoiding or minimizing activities such as drainage or wet flushing—the use of water to clean out large areas of dirt or insect infestations—as these are important, invertebrate-rich feeding areas, particularly for curlew chicks. The curlew will avoid nesting and feeding in areas close to tall trees and shrubs, so these should either be removed or heavily pruned.

Osprey
(Pandion haliaetus)

The osprey is among the most majestic raptors. It is commonly associated with water, as the seashores, bays, lakes and rivers are its habitats. The osprey often makes its nest on an elevated perch high above the waters, (e.g., a mast, lightpost or high cliff). An osprey is identifiable by its unique appearance consisting of a dark band from its beak to the brown on its back. It has white feathers all along its breast, crown, forehead and belly, and typically has bright yellow eyes. The female tends to be larger and has dark chest bands that are absent in the male. From a distance, the osprey has a distinct wing shape with a kink at the elbow and a rounded tip, but because the wing posture is different from other raptors, it sometimes is mistaken for a gull. However, its size, soaring pattern and distinctive call distinguish it from other, more common birds. (Kirshbaum and Watkins 2000) ("Osprey *Pandion haliaetus* Identification Tips" 2007)

The osprey is a fairly large raptor with long, thin wings, growing up to 55 centimeters in length, with a wingspan of 150–180 centimeters. It weighs from one and two kilograms and can carry food almost one-third of its weight. Another distinct feature is its talons—unlike other diurnal raptors, the outer toe of an osprey is reversible, making it easier to ensure a firm hold on captured prey, which it typically carries head first to reduce wind drag. Additionally, the pads of its feet feature modified, spiky scales that help it grip slippery fish. (Unitt 2000)

Local mullet are a favored prey of the osprey. When flying, the osprey typically holds its prey parallel with its body.

The osprey is a seasonal bird that migrates annually in the Western hemisphere from as far north as Alaska to as far south as Argentina. It typically spends the summer, its breeding season, in the north and migrates towards the south during the winter. In San Diego the osprey is a year-round inhabitant because of the temperate climate. According to a survey of North and Central San Diego Bay, the osprey is most frequently seen during the months of September and October. (Unitt 2000)

The breeding season of the osprey varies between populations. Typically, it reproduces in the summer when the temperatures are warmer. Nonmigratory populations like those that inhabit the San Diego area lay most of their eggs between the months of December and March and may nest into the summer. Migratory populations lay their eggs in April and May. In either case, the female will lay between two to four eggs per nest, over the course of several days. Both the male and female will incubate the eggs, which hatch in the order in which they were laid after 32 to 43 days. (Kirshbaum and Watkins 2000)

The osprey population in San Diego has rebounded since the ban of DDT in the 1970s. Nesting pairs have chosen locations in various spots all around San Diego County, always staying relatively close to the water. In 1999, a pair of ospreys began to nest at Scripps Ranch High School and had annual success for many years thereafter. Another nesting pair at Mesa College had some questionable success; one of the birds became entangled in a wooden fixture built on a lightpost and was rescued by the wildlife rehabilitation center in Ramona. (Kucher 2006). Other nesting sites include a boat mast in Mission Bay, San Diego State University, the east corner of North Island Naval Air Station and the Torrey Pines State Reserve. All nests have been less than four kilometers from a local water source for the osprey's dietary needs. (Unitt 2000)

The osprey can be quite fastidious about its nesting site and may delay breeding in order to find a suitable location. Because in San Diego there are no large tree stumps, the osprey's natural and native nest-site choice, and in order to keep its eggs away from predators, it selects high structures close to food sources. These structures are usually man-made, such as channel markers or lightposts. In addition, an osprey may return to its nest for many years, adding more material each year. Older nests can weigh hundreds of kilograms. (Unitt 2000) ("Osprey: General Info." 2007)

The osprey's dietary patterns

are another defining feature. Unlike most raptors, this consists mostly of fish, the types often dependent on the area where the osprey lives. The osprey is usually an opportunistic eater and will consume whatever fish it is able to catch, so if the catch is poor, the osprey will relocate. Therefore, the relative health of a lake or river can be determined by the presence of a substantial osprey population. Generally, the presence of osprey represents a healthy aquatic environment. (Kirshbaum and Watkins 2000)

The osprey normally gets its food from shallow waters, and it is spectacular when diving. Once the prey is spotted, the osprey swoops down for the kill. Moments before plunging into the water, the raptor pulls its wings in close and stretches them back. When it is roughly a meter above the water, the bird projects its talons forward to catch the fish. After plunging, it surfaces with an incredible force, pulling its weight and that of the prey up out of the water. Then it shakes its body and pulls one leg forward to hold the fish parallel to its body in flight. There have been cases where an osprey has tried to catch a fish that is too big for it to carry, so the bird is forced to let the prey go in order to prevent itself from drowning. ("Osprey: General Info." 2007)

Even the origin of the name "osprey" holds unique qualities. Its scientific name, *Pandion haliaetus*, is derived from an old myth about the Athenian King Pandion whose two daughters were turned into raptors. *Haliaetus* comes from the Greek words *halos* (sea) and *aetos* (eagle). Its common name, osprey, is said to have come from the Latin *ossifragus*, meaning bone breaker. The osprey's connection with oceans, rivers and lakes, along with its majestic and fierce nature, makes it the irrefutable king of our local waters. ("Osprey *Pandion haliaetus* Identification Tips" 2007)

DECLINE OF THE
Osprey

Once abundant near water systems throughout the country, osprey populations nearly collapsed in the 1950s and 1960s. John James Audubon once wrote of the osprey, "Fish Hawks are very plentiful on the coast … where I have seen upwards of fifty of their nests in the course of a day's walk." This contrasts to today's numbers, making concerns about the osprey's status apparent.

Like many other birds, the osprey population was significantly damaged by DDT because of a direct effect on its population in the form of egg-shell thinning. Similarly to other impacted species, the eggshells became so thin that they could be easily crushed during incubation. The osprey population nearly collapsed and nesting all but disappeared in Southern California. Then, in 1972, after the harmful effects of DDT to osprey and many other species were recognized nationwide, DDT use in the United States was banned completely and efforts to aid the osprey and other species affected by the pesticide's use were enacted. Although the ban helped the bird to rebound, other equally detrimental causes would keep the osprey from returning to its original population numbers before human impact. For more information on DDT, see "An Inside Look" on page 76. ("Osprey" 2007)

In addition to pesticides, direct human disturbance has had a great impact on osprey populations. Historically, it was often killed by fishermen who not only believed it was competing for fish but valued its eggs. Although osprey hunting is now illegal because of the bird's conservation status, its once large numbers and numerous nesting locations made it an easy target for poachers. Its large, ostentatious nests, as well as its tendency to build on elevated man-made structures, was somewhat of a downfall. This not only made the osprey easily visible, but it was found closer to urban sites than most other birds. With this knowledge, hunters were able to locate its whereabouts and move in for the kill.

Habitat loss has proved to be just as detrimental as DDT and hunting for the osprey. Throughout California, as well as the world, the

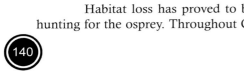

bird has learned to cope with many of the new structures humans have built on its natural habitat. Active nests were once removed from channel markers and buoys, while other nests were completely demolished to clear areas for coastal residences. Today, habitat loss still poses a threat to all raptors; some ospreys have failed to adapt to their new urban landscape and injure themselves. Although, in general, the osprey has now adjusted to nesting on man-made structures when there are no natural ones, additional development of its breeding areas will only pose a greater threat. (Unitt 2000)

RECOVERY OF THE
Osprey

Since the decline of the osprey population, mostly due to DDT exposure, there have been many attempts by humans to restore the "King of the Sea." Though briefly on the federally endangered species list in 1976, the osprey has been categorized as a Species of Special Concern since the 1990s. Current osprey populations would not exist today without the courageous advocacy, wisdom and perseverance of Rachel Carson. Her thought-provoking 1962 book, *Silent Spring*, promoted an environmental movement, which fought common pesticides that were destroying bird populations worldwide. The osprey served as a "poster bird" of the Carson classic and the environmental movement that followed, which, in turn, dramatically increased public awareness and the resurgence of the osprey in our communities.

In San Diego County in particular, there have been many instances of direct human involvement to restore the native osprey population. In 1997, a pair of ospreys nested in San Diego for the first time since 1912. Since then, the community has recognized the importance of maintaining local osprey populations. Concerned citizens, local bay stewards and even large companies have joined the cause. One such case is Dixieline Lumber, a leading San Diego lumberyard, which has helped build numerous osprey nest sites throughout the county, starting in their

own backyard with some lightposts. Many of these nest sites are now occupied by the osprey and promote its breeding around San Diego Bay.

The media has also played an important role in osprey recovery. In June 2006, the *San Diego Union-Tribune* wrote a story about a fledgling osprey whose wing became entangled in a wooden fixture built on a lightpost at Mesa College. Despite an effort to rehabilitate the young bird, it died because of the injuries sustained. (Gibbons 2006) Though its death was quite saddening, the media coverage of the bird brought awareness and community support. Efforts to restore osprey populations in San Diego have increased because of the steps local citizens and companies have been taking to reverse damages that threaten local species. (Kucher 2006)

In 2006, the Port of San Diego Port established a $5-million environmental fund to be used for "research, grants and programs aimed at restoring habitats and nurturing San Diego's waterways and shoreline." (Magee 2006) As part of their recovery efforts, the Port set aside approximately $40,000 to construct five osprey nest platforms around San Diego Bay, each consisting of a pole between 3 to 5 meters in height, topped with a large platform. The platforms were built in Pepper Park, the Chula Vista Wildlife Reserve, Emory Cove, Spanish Landing and Shelter Island. Construction of the nesting platforms began in mid-April 2007 with hopes to track the number of ospreys that nest and the eggs that hatch as soon as possible. This has been a great asset to the osprey, further improving the health and longevity of its population and its habitat. (Magee 2007)

Although there have been many instances of preserving the osprey population, there is still much more that needs to be done. The only way this can happen is through education, preservation and community involvement. It is important to understand the need to ensure that our actions do not endanger animals, especially those with fragile populations. In order to ensure a good quality of life for birds-of-prey like the osprey, the destruction caused by pollution of water and land must be reversed and citizens need to be mindful of their surroundings. These impacts (often occurring far away from the coast) can be significant on the overall quality of life, not just for the osprey, but for all species.

Peregrine Falcon
(Falco peregrinus)

The peregrine falcon is nature's fastest bird and a beautiful raptor. Along with the osprey, the peregrine is one of the most widely distributed birds of prey in the world. Its species name is appropriate—from the Latin *peregrinus* meaning "wanderer" or "traveler." Some subspecies of the falcon have been known to migrate longer distances than any North American bird. Its adaptability to a variety of habitats have allowed it to live on all continents, with the exception of Antarctica, the three peregrine falcon subspecies that inhabit North America are the American (*Falco peregrinus anatum*), Arctic (*Falco peregrinus tundrius*), and Peale's (*Falco peregrinus pealei*). The American peregrine falcon can be found soaring the skies over San Diego County. ("Peregrine Falcon[2]" 2007) ("Peregrine Falcon[4]" 2006)

Peregrine falcons are well-known for their striking features. Adults have a dark crown with two, thick "mustache" markings and white cheek patches, making the bird appear to be wearing a hood. The upperparts of the adult are bluish-gray and the underparts, including its chin, are whitish in color. It has barring on its belly and leg feathers, which are most notable when the bird is in flight. The peregrine has large, yellow feet and a short, hooked and notched beak. Its long and slender

wings contribute to its aerodynamic skill. When perched, its wing tips almost reach the end of its tail. The juvenile peregrine looks similar to the adult, except with darker underparts and more brown in its feathers. (Brown 2006) ("Peregrine Falcon" 2007) ("Raptor Identification" 2007)

The peregrine is a medium-sized raptor, roughly the size of a large crow. It ranges in length from 38 to 53 centimeters, with a wingspan of about 100 centimeters. Male peregrines are known as "tiercels," meaning "one-third," because they are about 30% smaller than the female. The male average weight is 0.7 kilogram, while the female weighs about 1 kilogram. The peregrine's flight pattern is direct with rapid wing beats. ("Peregrine Falcon[3]" 2007) ("Raptor Identification" 2007) (White, et al. 2007)

The hunting prowess of the peregrine is defined by its power and incredible speed. When stooping (diving for prey) from a high altitude, the peregrine may reach speeds more than 400 kilometers per hour. Its main hunting tactic is to soar to a great height above other avian prey, then dive almost vertically to ambush the prey in midair. It will also pursue prey in a low and fast flight, or attack passing birds from a perch. During peregrine falcon studies in San Diego, the bird was seen stooping over shorebirds along the water's edge. ("Peregrine Falcon[3]" 2007) ("Peregrine Falcon" 2007)

The diet of a peregrine falcon primarily consists of small- to medium-sized birds. These include starlings, pigeons, blackbirds, jays and various shorebirds and waterfowl. On rare occasions, the peregrine will also eat small mammals, reptiles, fish or insects. The female can capture larger prey because it is larger than the male. ("Peregrine Falcon" 2007) (Unitt 2004).

Generally, the bird is found in open habitats, from coastal regions to high mountains. It usually nests on tall cliffs with a nearby water source, an open area and abundant supply of prey. It also uses the high cliff bluffs for perching and hunting. The peregrine will nest primarily on ledges, caves or small holes that provide weather protection. It typically breeds in a coastal or riparian environment. However, because of habitat loss and the bird's adaptability, it is now frequently found living and breeding in many urban areas throughout North America. In the San Diego region, peregrine falcons have been nesting around Sunset Cliffs, Torrey Pines,

and Black's Beach. ("Peregrine Falcon" 2007) (Sooter 2007)

The peregrine will often make a nest, or eyrie, from a small depression in dirt and line it with debris. If the environment is hospitable, these nesting sites are typically used for many years. The falcon has also been known to take over the nests of other birds rather than building its own. When nesting on a building, the peregine will construct a nest out of twigs and dirt. The falcon reaches breeding maturity at the age of two. In the San Diego area, it breeds from early March to late August.

Peregrines skillfully use their hooked bill to rip away feathers and skin to expose the flesh of their prey that mostly consists of small birds.

The male will usually arrive at a nesting site and perform a series of aerobatics to attract a mate. The female will lay between three to seven eggs at two- to three-day intervals. Incubation is carried out by both parents and lasts approximately 35 days. Young peregrines fledge within five weeks of hatching. While caring for young, peregrine parents can become aggressive and defensive about their territory. Peregrine parents were observed stooping on airborne offenders as large as California brown pelicans, ospreys, and even hang gliders. Such intruders flying near the peregrine nest were not only chased but frequently attacked by the outraged parents. ("Peregrine Falcon[2]" 2007)

Once the young fledge, the parents assist in training their offspring to fly, stoop and kill. These hunting lessons are important in preparing fledglings to survive on their own.

This is an example of a peregrine falcon in its famous hunting "stoop."

DECLINE OF THE
Peregrine Falcon

Early pressures on the peregrine population were suspected to be connected to overdevelopment, egg harvesting, and misguided persecution from those who believed raptors were detrimental to farming and ranching. However, nothing reduced the peregrine falcon numbers more than DDT. When the pesticide was in wide use, food-based doses of the chemical accumulated in the bird's fatty tissue and, as with other avian species, eggshell thinning also occurred. The raptor's death rate soon exceeded the birth rate, and the population declined rapidly. By 1970, California's peregrine population had been reduced by more than 95%, with only two known nesting pairs, neither of which resided in San Diego. The effects of DDT were so great that it landed the peregrine a spot on the endangered species list in 1969. ("Peregrine Falcons and DDT" 2007) (Pavelka 1990)

Even when DDT use diminished, its effects were perpetuated through residual DDT and DDE (dichlorodiphenyldichloroethylene), a metabolite of the pesticide that is still present in our environment today. As a result, these contaminants continue to pose a serious threat to some populations of peregrine falcons well after the ban on the chemical. ("Peregrine Falcons and DDT" 2007) For more information on DDT, see "An Inside Look: DDT in the Environment" on page 76.

In addition to DDT, habitat loss has cost the lives of many peregrines. The falcon is often forced to adapt to an urban environment. Though many have learned to cope, others fall victim to dangers like automobiles and human disturbance. Another urban peril is when the peregrine dives or stoops, it risks crashing into fences and walls that are not as forgiving as nature's branches and bushes. The survival rate for the peregrine that lives in an urban setting is about 10% less than for the bird living in natural surroundings.

One local example in which urban life displaced and injured the peregrine population occurred in 1989. A pair of falcons nested on the San Diego-Coronado Bay Bridge and hatched four chicks: three males and one female. Two of the males fledged successfully; the third, however, flew into a nesting tower and fell to the concrete below. He survived, although with a broken wing. The fourth nestling, the only female, was hit by the San Diego Trolley while learning how to fledge. (Pavelka 1990)

The same nesting pair returned the following year and hatched three more falcons: one male and two females. The male died after he was hit by a car on the bridge. Both the remaining females fell to the ground from the nest or during fledging. They were later taken to the University of California, Santa Cruz Predatory Bird Research Group's (SCPBRG) facility, where they were raised in captivity and eventually released in Yosemite National Park. One of the females became completely independent, but the other was killed by a golden eagle shortly after being released. The survival rate of first-year raptors is naturally low for any species in the wild, and the hazards created by human interference only makes it more difficult for these treasured animals to survive. (Pavelka 1990)

Sadly, in 2007 it was reported that hobbyists who breed and fly the roller pigeon, a non-native species, had been shooting and trapping peregrines and other raptors because they were considered a danger to their exotic birds. These are only a few examples of urbanization threats to the falcon. With its unique nesting requirements, dietary needs and perilous fledging, it is nearly impossible to safeguard the species from human-induced dangers. We must hope that the resilience of nature and the falcon's adaptive abilities will prevail, and someday restore its population to a more sustainable state.

Several months after fledging along the cliffs of Black's Beach, the juvenile peregrine, "Edge," was observed making a pass over the J Street salt marsh in search of food.

RECOVERY OF THE
Peregrine Falcon

In 1970, the peregrine falcon was listed as endangered under the Endangered Species Conservation Act of 1969 (the federal law preceding the Endangered Species Act [ESA]). With the status of endangered, more attention and support was given to research of the peregrine's precipitous decline. The U.S. Fish and Wildlife Service (USFWS) established peregrine falcon recovery teams, composed of federal, state and local biologists, to research and recommend actions to recover peregrine populations in the United States.

In 1972, the Environmental Protection Agency banned DDT for most uses in the United States. ("Peregrine Falcon (*Falco peregrinus*)" 2006) With the banning of the pesticide, various environmental conservation organizations were established to aid in the recovery of the peregrine and other birds of prey affected by its use. Established in 1970 by Dr. Tom J. Cade, one of the first scientists who researched the effects of DDT, the Peregrine Fund was created to bring back the species from near extinction. Dr. Cade, a professor at Cornell University, used captive breeding, species reintroduction and public outreach as the fund's primary techniques to restore the raptor's population. Over the years, the fund has expanded its efforts to more than 24 species of raptors and several nonraptor species. Today, the fund exists solely because of private donations and is currently working on more than a dozen different national and international conservation projects. ("Zoological Society's Highest Honor ..." 2006) (Lincer 1975)

Like the Peregrine Fund, the SCPBRG has contributed largely to the recovery of peregrines. The SCPBRG dates back to the 1970s after Dr. Jim Roush, a Santa Cruz orthopedic veterinarian, heard of Cade's captive breeding of peregrines at Cornell University. Inspired by Cade's work, Roush decided that similar efforts were needed on the West Coast. The group spent much of its time attempting to track down and locate the last remaining peregrine nest in California. In 1975, the SCPBRG began its breeding, conservation and management of falcons. From then through the mid-1990s, a large-scale release effort of captive-bred and captive-hatched peregrines was conducted. The SCPBRG no longer breeds peregrines, focusing instead on field projects related to predatory birds in California. (Walton 2006)

While the DDT ban had a large positive impact on the peregrine population, the recovery of the species is mainly due to the efforts of conservation organizations, such as the Peregrine Fund and SCPBRG. Today, because of captive breeding and management programs, the peregrine population has reached an historic high. The bird was removed from the federal endangered species list in 1999, marking one of the most successful recovery efforts since the creation of the ESA. Then-U.S. Secretary of the Interior Bruce Babbitt described the event as, "a spectacular moment for one of America's most noble birds and the law that helped make its recovery possible." In 2006, the SCPBRG conducted a statewide census of the species in California and documented more than 200 areas in which the bird was residing.

Various conservation organizations are still working to maintain peregrine populations. In the San Diego region, wildlife photographer and citizen scientist Will Sooter has recently established a peregrine falcon conservation group. As of this writing, the falcon is listed as a "species of concern" and is subject to monitoring. In 2003, the USFWS began surveying the peregrine after it had been delisted from perilous status. The USFWS monitoring efforts, in partnership with SCPBRG, will continue until 2015, after which they will be conducted every three years. Through the efforts of concerned individuals and organizations, the peregrine falcon remains a perfect example of why active wildlife conservation is important in San Diego and around the world.

The Chula Vista Nature Center cares for a captive peregrine with a broken wing. The bird's high speeds while hunting can lead to a number of serious injuries. Living in an urban environment only increases the possibility of such injuries.

Will Sooter

Will Sooter, a local naturalist and wildlife photographer, has had a variety of careers, ranging from sea-going field biologist to the founder of a retained executive search firm here in San Diego. Nowadays he conducts field observations and keeps a yearly chronology of peregrine falcons along our coastline. The nesting and monitoring data he compiles are contributed to the University of California, Santa Cruz Predatory Bird Research Group.

As we sat down with Mr. Sooter on a bluff overlooking Black's Beach in La Jolla, gray clouds filled the sky and a cool wind blew in off the ocean. We sat just above the site where, only three weeks earlier, a UC field researcher and associate of Mr. Sooter's had rappelled down the bluff to a peregrine eyrie (nest) to place ID bands on three sixteen-day-old young chicks.

During our previous visit to Torrey Pines State Beach with Mr. Sooter we had not seen any peregrines along the bluffs, but this time at a different location we were fortunate as a young male fledgling, who had been named "Edge," flew by just before we began our interview. Some five minutes later, movement in the sky caught Mr. Sooter's eye again and he stopped us to point out another young peregrine fledgling flying out from the bluffs and going into a dive to attack a group of pelicans flying over the shoreline.

Student Researcher (SR): What interested you about the peregrine falcon, and what do you find most remarkable about the bird?

Will Sooter: Five years ago, I used to run on the beaches from Del Mar down to Torrey Pines, and while running one day, I observed a peregrine falcon perched on the bluffs. At the time, I had taken a break from working at my search firm and was looking to pursue a different path. This beautiful but still endangered peregrine falcon caught my attention and the thought came to me that I should try to observe and photo-document the peregrine's behavior on a daily basis. I thought it might be a good way for me to raise awareness about conservation issues and the need to preserve what little wildlife habitat we have left on the San Diego coastline. I had no idea that this was going to become a "project" and the ambassador for my new mission. By observing the peregrine on a daily basis, my life was changed forever. After my first season in the field observing him, I named this tiercel (male) falcon "Stretch." I found out that he was not a native *anatum* peregrine, but a visiting migratory *tundrius*. He would migrate to Torrey Pines the same week every October

The peregrine falcons of San Diego are fortunate to have the likes of avid naturalist Will Sooter, who works towards protecting the species.

and then depart for his home in the northern latitudes in late March or early April. For the next four years I followed him daily with my binoculars and camera gear. One sad day in 2007 I found him dead on his cliff-top perch. By that time, I had received a tip from San Diego City lifeguards about more peregrines in La Jolla. Together we discovered a mating pair of native (*anatum*) peregrines here on the bluffs at Black's Beach.

To answer the second part of your question, I find peregrine falcons fascinating and "addictive" birds of prey. They are at the high end of the food chain; they have unique physical and aerodynamic characteristics that allow them to dive at high speeds to pursue their prey (or other birds) in mid air. In fact, they are the fastest bird on the planet; they've been clocked at more than 240 miles [400 kilometers] per hour in their dives that we call "stoops." You can view videos of this on my website (www.sharpeyesonline.com).

Peregrine falcons had a rough go of it in the 1950s and 1960s; they almost became extinct due to widespread use of the pesticide DDT. It got into their food chain and lowered their reproduction and survival rates. By 1970 the peregrines were listed as an endangered species and in 1972 DDT was banned for use in North America. Now, thanks to heroic recovery efforts of many concerned scientists and individuals, the peregrine has made a remarkable comeback. If it starts disappearing again from our local habitat, then we'll know something is not right with our ecosystem.

SR: DDT had an incredibly detrimental impact on the peregrine population; what are some modern-day "DDTs" that limit the bird's recovery and success locally?

Sooter: From my observations, I see several potential modern-day equivalents to DDT. Recent studies by scientists in California have found many peregrine falcons that are now contaminated with extremely high levels of a toxic chemical known as polybrominated diphenyl ethers (PBDEs). These are industrial chemicals used as flame retardants on electronics and furniture. They eventually migrate outdoors where they pollute the urban environment. The scientists have found that peregrines are being contaminated by this chemical when eating urban pigeons that inhabit our city streets. This is very disturbing news and I see these PBDEs as being both an unnecessary and dangerous threat to our now recovering local peregrine falcon population.

Another threat to the peregrine is urban street runoff and water pollution in San Diego County. On these cliffs where we are sitting, there are several canyons that have intermittent streams running down them toward the ocean. Polluted and sometimes chemical-laden runoff from the nearby business facilities, public golf course and large estates situated on top of these cliffs eventually ends up in these streams and then flows out onto the beach and into the ocean. The runoff water also forms vernal pools up in the canyons and shallow pools at the base of the cliffs, where our local peregrine falcons bathe and drink. In 2007 I watched "Stretch," the migratory tundra peregrine, die within 24 hours of bathing and drinking in one of these canyon pools. He could have died from many things, but just knowing that he was bathing in polluted runoff left me with a suspicious and uneasy feeling.

A third threat as I see it, is the continued development along our coastline that only adds to more water pollution and habitat destruction. We have so little open space left along our coast, it would be a shame to squeeze peregrine falcons out of their historic wild habitat. Along these La Jolla bluffs we have one other unique potential threat to peregrines—the paragliders who launch and fly out from the Torrey Pines Glider Port and share the same airspace. During nesting and brooding season these paragliders fly into the canyon and often very close to nesting sites. In fact, when this happens the adult peregrine will defend its territory by flying out and attacking the paragliders. I have photo-documented this attack on paragliders dozens of times. It is disturbing as it causes undue

stress for the bird during the critical and sensitive nesting season. For the time being, it seems that the peregrines have adapted somewhat to the paraglider presence, but I worry that an increase in such flight activity near their eyrie could agitate the peregrines enough to cause nest failure and thus force them out of their habitat.

SR: What have conservation groups and organizations done to restore the peregrine population? What about specific individuals?

What can individuals now do to help?

Sooter: The Peregrine Fund is a private non-profit organization that works nationally and internationally to conserve birds of prey in nature. It was founded in the 1970s by Dr. Tom Cade and began a successful captive breeding program to restore the peregrine population in the United States. They, along with other organizations like the UC Santa Cruz Predatory Bird Research Group, were really responsible for bringing the per-

"Sid" and "Nancy" perform a midair food exchange along the bluffs of Torrey Pines. This amazing behavior is another one of the falcon's unique characteristics.

egrine's population back from the brink of extinction and to its healthy level in the United States today.

Specific individuals that I personally know who have greatly contributed to the recovery of the peregrine falcon throughout the United States and who have encouraged me in my work here in San Diego include: Scott Francis, Project Coordinator for the Santa Cruz Predatory Bird Research Group; Dr. Daniel Brimm, scientist and founder of the Peregrine Recovery Council, La Jolla; Dr. Clayton White at Brigham Young University; Dr. Jim Enderson, Professor Emeritus of Biology at Colorado State; and Dr. Tom Cade, Founder/Director of the Peregrine Fund and Professor Emeritus of Ornithology at Cornell University. We really owe a great debt of gratitude to these individuals and so many more like them who contributed to restoring our peregrine falcon population both in California and throughout the United States.

Individuals who would like to become involved can volunteer time to local conservation groups or donate money to organizations like the Peregrine Fund who do specific research on raptors and try to protect them. People who would like to support my local peregrine falcon monitoring and conservation efforts may do so by visiting my website (www.sharpeyesonline.com) or contacting me.

As you know, we have very little open space left in coastal San Diego. There are just a few isolated, undeveloped places—what I call "islands of open space"—where peregrines can establish themselves in truly historic wild habitat. The cliffs at Black's Beach and Torrey Pines are unique in that they are currently isolated and protected from further development. These bluffs provide a perfect habitat for peregrines, who like high perches for nesting and hunting. There was a time when this was their space and their space alone. That is no longer the case. If further development is allowed on the bluff tops or in the canyons, and sea walls are put up to protect the development, there will be less habitat left for peregrines and other wildlife along the coast. San Diego County citizens can do a lot more to protect our existing natural resources by speaking out against over development along the coast and the new trend toward commercialization of our State Park Natural Reserves, as you see going on at Torrey Pines.

SR: Could you describe some of the behavior that you've noticed in the peregrines? You're out here all the time, and surely that's a truly different perspective. To get so close to these birds, you have to notice a fair bit about their habits.

Sooter: I would be glad to. I get so much pleasure from spending hour upon hour observing these magnificent birds of prey. Imagine yourself in the Serengeti Plains of Africa and you're out there day-in/day-out watching lions or cheetahs chase down their prey. It's an incredible sight to watch a peregrine falcon flying out over the ocean, using its amazing binocular eyesight to spot its prey at more than a half-mile away, and then going into a steep dive (stoop) sometimes at more than 200 miles [320 kilometers] per hour to capture its prey in mid air. That is a sight that not many people get to see because they don't have the patience, time or ability to keep track of a peregrine in a stoop. That means being here all the time—at least four to eight hours a day—to get in sync with the individual peregrine's daily habits and behavior. Not only do you need to know your subject but you also need a good pair of binoculars, and the ability to track the bird at some pretty good distances and in high-speed flight. Sometimes the male (tiercel) peregrine will tandem hunt and stoop with the female (falcon) in going after their prey. A successful peregrine stoop on its prey is a sight to behold.

During nesting season, there is a specific type of behavior that you don't see during the rest of the year. It is called a food exchange. This is the transfer of a prey bird in mid flight between the tiercel and the falcon. While the female is brooding her eggs or young in the eyrie, the tiercel peregrine will do the hunting and provide food to her and their young. After a successful hunt where the tiercel captures a prey bird, he will sometimes return to the bluffs near the eyrie where he calls out vocally to the female. She will then fly out to meet the tiercel who, in the meantime, switches the prey bird from his beak to his talons and then in a split second the female inverts herself under her mate as he transfers the prey bird from his beak to her upturned talons in mid air. It is a spectacular feat of aerial acrobatics! The actual exchange happens fast—in three to five seconds—and to observe it you have to be ready. I've been fortunate enough to observe many food exchanges over the last several years, literally standing in the same spot for several hours anticipating one so I could attempt to photo-document it. To me, as field naturalist and wildlife photographer, capturing photos of food exchanges are the "holy grail" of peregrine falcon photography.

Northern Harrier

(*Circus cyaneus*)

The Northern harrier, also called the marsh hawk, is a symbol of the wetlands, marshes and grasslands. The bird's slow and somewhat erratic flight distinguishes it from other birds of prey as it frequently deviates from a direct course, dropping almost to the ground and then soaring up again in search of food.

The California Department of Fish and Game named the Northern harrier as a species of special concern in 1978. The harrier is a raptor belonging to the Accipitridae family and is found throughout North America, especially in Southern California, Texas, Northern Virginia and New England, in protected marsh and lowland areas. It migrates from late February to early March to most parts of the United States, Central America and the Caribbean. Because its winter migration covers such a wide area, the spring migration can vary in range, depending on where the bird originated. (Remsen 1978)

The ideal habitat of the Northern harrier consists of vast marshes, free from disturbances and cloaked with tall, dense vegetation. This vegetation usually includes broad-leaved plants, low shrubs, grasses and sedges. The harrier may also enjoy open habitats: old crop fields, meadows and prairies, and its preferred nesting site is grassland or ruderal vegetation areas where the ground is dry. A pair of harriers require at least 250 acres of habitat for an adequate hunting area. ("The Northern Harrier: *Circus cyaneus*" 2007)

This beautiful creature can be recognized easily by its slim body and long, feathered tail. The male has a white breast, a pale gray back, and black wing tips. The female has a buff-colored underbelly with brown streaks and a brown back. The wingspan for both the male and female harrier spreads to an average of 110 centimeters. A unique feature of the bird is its owl-like facial disk, providing it with an excellent hunting advantage; it uses this disk when flying low to capture the sounds of its prey. The harrier can be identified by its low, characteristic flight and heavy wing beats. ("*Circus cyaneus* Linneaus: Northern Harrier" 2008)

Because the Northern harrier has keen hearing and eyesight, it can hunt a wide variety of prey, mostly small mammals, birds and insects, such as grasshoppers. The smaller animals (mice and birds), are hunted in saltwater marshes during the winter when grasshoppers and other insects are scarce. The harrier combines the use of strong talons, a powerful, hooked bill, and low-flight swoop to capture its prey. ("The Northern Harrier: *Circus cyaneus*" 2007)

DECLINE OF THE
Northern Harrier

The Northern harrier population in California began to decline as early as the 1940s, mainly because of reproductive failure caused by the harmful pesticide DDT. The widespread use of this pesticide during the mid-twentieth century continued to heavily affect the harrier until the federal ban of DDT in 1972. (Ehrlich, Dobkin, and Wheye[1] 1988) For more information on DDT, see "An Inside Look: DDT in the Environment" on page 76.

The harrier's numbers were scarce enough to merit placement from 1972 to 1986 on the National Audubon Society's Blue List of Imperiled Species, an early warning list of threatened North American bird species that had impending or ongoing declines. (Ehrlich, Dobkin, and Wheye[2] 1988)

Despite the ban of DDT, the harrier still remains imperiled because of the loss of vital habitats. With the onset of urban and industrial development in coastal areas, marshlands and grasslands

The continual encroachment of the Northern harrier's natural habitat has resulted in population declines.

have been in steady decline. Unfortunately, because of their great dependence on these habitats for breeding and feeding, the harrier population has followed a similar pattern. ("Bay-Delta Region River Report: Northern Harrier" 2008) ("Northern Harrier" 2008) ("Northern Harrier Fact Sheet" 2007)

Overgrazing is another issue. When plants living in the harrier's habitat are exposed to long periods of grazing, the level of vegetation deteriorates, affecting the ecosystem of the entire area and, eventually, the bird itself. In the end, despite the harrier's ability to adapt, its numbers are decreasing globally. ("Northern Harrier Fact Sheet" 2007) ("Northern Harrier" 2008)

RECOVERY OF THE
Northern Harrier

Despite the population scare of the 1970s and 1980s, the Northern harrier is now populous enough to be rated as a species of "Least Concern" by the International Union for Conservation of Nature. However, it is protected by the federal Migratory Bird Treaty Act. ("The Northern Harrier: *Circus cyaneus*" 2007)

According to the California Bay-Delta Region Stanislaus River Report, keeping habitats free from agricultural uses is important for the conservation of the Northern harrier population. The Northern America Wetland Conservation Act and the Freshwater Wetland Protection Act also work to preserve the bird's already dwindling wetland habitats and restore damaged regions so it may flourish. As the harrier relies on wetland health and availability for its survival, protecting these will in turn help protect the species and allow it to repopulate. ("Bay-Delta Region River Report: Northern Harrier" 2008) (Unitt 2004)

One area of San Diego County that appears to support the bird's recovery is the Tijuana River Estuary. In 2002, there were 13 pairs of harriers in the Border Field State Park. (Wildlife Research Institute 2004) However, the recovery of the Northern harrier may come with a price for other endangered species in that region, as it has been observed to prey upon the light-footed clapper rail and the California least tern, two of the rarest bird species in the area. (Unitt 2004)

During spring 2007, five Northern harrier nests were documented and additional nests were suspected of having been established in the Tijuana Estuary.

Willets take flight as the Northern harrier hunts along the coastal salt marsh.

Burrowing Owl
(*Athene cunicularia*)

The burrowing owl is a small ground owl mainly found throughout Western North America and as far south as Argentina. It is a unique member of the owl family because of its crepuscular lifestyle and underground nesting behavior. The owl prefers to make its nests in the abandoned homes of ground-nesting mammals, but it is not uncommon to see it dig its own burrows when necessary. It also readily uses man-made, artificial burrows that have been constructed especially for its use.

From a distance, the burrowing owl looks like a stout bird. Its square frame and large head gives it a typical owl appearance. However, on closer examination, its size shows how fragile a creature it truly is—only between 20 to 28 centimeters tall and weighing around 180 grams, it is one of the smallest owls in Southern California. It has a rounded head with flat ears, piercing yellow eyes, furrowed white eyebrow and long legs. Its underbelly is snow white with small brown dots while the rest of its body is various shades of light to dark brown, often with white spots. The male is usually lighter colored than the female, which may help to differentiate the sexes. ("Owl Biology" 2007)

When searching for a place in which to lay its eggs or simply a home, not just any unoccupied hole in the ground will do. The burrow must be large enough for entry and the immediate surrounding area must be clear of vegetation to permit an easy lookout for ground predators. The burrowing owl usually makes its home in grasslands, flatlands, deserts and prairies. The burrow walls may be lined with manure, which has several benefits: a soft surface for laying its eggs; absorption of water to prevent flooding in heavy rain; providing insulation from both hot and cold temperatures; and attraction of a variety of insects for food. The burrows generally have one or two chambers, one of which serves as an area for food storage and another for laying eggs and sleeping. Storing food within the burrow is not uncommon and many food items can create an impressive cache. (Wong 2004)

When a female is ready to nest, a two-week mating period is typical before she settles and lays her eggs. The female incubates the eggs for 28 days before they hatch. During this time, the male hunts and gathers food for both of them and, if enough is available, he will also save some for the young chicks. The owl hatches with its eyes closed, but after a few days, they open and it begins to explore the area in and around the nest. By three weeks, it is able to wander freely but still unable to fly. At four to five weeks, it can fly and learns to catch its own food, such as insects and small rodents. At six weeks of age, the owl starts to become independent but still relies on its parents before it leaves the nesting area. Approximately one-half of the surviving birds at six weeks make it through the first year of life. Mortalities include a variety of man-induced and natural conditions such as lack of food, unfavorable climate, uninhabitable homes and predator attacks. The typical lifespan of the burrowing owl is eight and one-half years. (Restani 2001)

The owl's diet varies depending on its specific environment, but largely consists of insects (mostly grasshoppers, scorpions, moths and beetles), rodents, other small mammals and some reptiles. They will also occasionally prey on some songbirds. When feeding, the owl uses its talons to hold down the food while ripping it apart with its hook-shaped beak. When the prey is too large to be eaten all at once, the owl will store the food or share it with its mates. ("Owl Biology" 2006)

The northern populations of burrowing owls tend to migrate to other areas in the winter. The southern population owls, like those found in San Diego, usually do not migrate, but may change burrows once or twice annually. One San Diego population can be found in the Otay Mesa area, a flat, open region that is ideal because of its warm climate and lack of human interference. However, increasing development puts even this Otay Mesa population in jeopardy. (Unitt, 2004) (Gailband 2007)

DECLINE OF THE
Burrowing Owl

"When I first came to San Diego from New York State in 1921, a surprising discovery, among many interesting new bird experiences, was the occurrence of burrowing owls in well-settled parts of the city. On El Cajon Boulevard, which was a well-traveled thoroughfare even in those days, burrowing owls could often be seen perched on the sidewalk curb." (Abbott 1930) Over the past 50 years, the population of the owl has decreased drastically. It was once populous all throughout the West Coast of the United States and Canada. Back then, the little owl would use all the land that was available to it. It was once said by an ornithologist, that "burrowing owls stood on every little knoll" around San Diego. However, urbanization has led to the destruction of the owl's home, the primary cause of its endangerment.

In the 1970s, the burrowing owl population peaked, with more than 2,100 breeding pairs in San Diego and Orange Counties combined. However, within seven years, this population dropped to almost one-half of the previous count; and a few years later, there were only 300 breeding pairs left. By 2000, there were fewer than 12 burrowing owl breeding pairs in the two counties.

There are several reasons for such a large decrease in the population; the main one is the destruction of its subterranean dwellings and surroundings as buildings were constructed and more land developed. Although the owl is partially able to adapt to buildings and construction, it is nearly impossible for it to thrive. Open space is rare in San Diego and the continual land development makes it harder for the owl to nest and forage. The owl requires a specific type of vegetation—not too tall yet not too sparse—to enable it to hide from predators.

Pesticides and herbicides also cause both direct and indirect damage to the owl. Pesticides eliminate insects and rodents, its primary food source. Since the owl is known to scavenge dead prey, this makes it highly sensitive to secondary poisoning; and it is sometimes affected by chemicals used to control squirrel and rodent populations. Other toxins also cause problems for the owl—strychnine-coated grain, used to control ground squirrels, has been found to cause significant loss of body mass and a decrease in breeding success. (Sheffield 1997)

Extensive land use for new homes, industry and supporting infrastructure are major reasons for the owl's waning population. Many legal battles have taken place in recent years between developers attempting to build in the owl's nesting areas and local conservationists trying to preserve the fragile population. Humans often prioritize their needs over the needs of wildlife, and irresponsible decisions made in this mindset are the largest causes for the decline of the burrowing owl and other species.

Burrowing owls have seemingly been attacked from all sides, but fortunately they have the likes of Charles Gailband and the Chula Vista Nature Center on their side.

RECOVERY OF THE
Burrowing Owl

Since the decline of the burrowing owl population, several recovery programs in San Diego County have been implemented to reverse the negative trend of this specialized bird.

In order to help facilitate the recovery of the bird and its habitat, biologists and volunteers have installed man-made burrows to increase breeding and aid in predator avoidance. These artificial burrows are subterranean structures made out of drainage pipe and wooden boxes to simulate natural burrows in areas where there are none. Each burrow has two funnel-shaped entrances for easy escape in the event of an attack by a ground predator. The entrances are reduced in size to prevent larger predators from entering the breeding chambers. (Lincer 2005)

Volunteers with the World Resources Institute (WRI) built 52 man-made burrows in the Ramona Grasslands and were subsequently given permission to install artificial burrows in the Lower Otay Lake area. Hopefully, this will attract more owls and help restore their population. (Lincer 2005)

The U.S. Fish and Wildlife Service and the Sweetwater Authority are also conducting burrowing-owl habitat improvements and artificial burrow installation. A large area along the eastern shore of the Sweetwater Reservoir and a portion of the San Diego National Wildlife Refuge have been modified for the owl. The improvements have attracted both resident and migrant birds. Additionally, several captive-reared owl chicks from the Chula Vista Nature Center were released and tracked in the area in spring 2008. It is now home to several confirmed breeding pairs and several individuals.

A larger program that is greatly assisting the North American population is located in Canada and is appropriately named, "Operation Burrowing Owl." In that country, it is illegal to kill or disturb the owl and its home; and Alberta farmers have agreed to set aside more than 50,000 acres of grassland for the owl. The grassland would be cultivated where there is a history of burrowing owl habitat. Through this program, it is hoped to increase the food supply for owl pairs and the number of offspring they produce. The goal is to increase the owl population to at least 1,000 breeding pairs. ("Burrowing Owl: An Endangered Species" 2001)

Gray Whale
(Eschrichtius robustus)

Those fortunate enough to observe the gray whale would agree that it is one of the world's most unique and majestic mammals. It is the only member of the Eschrichtiidae family and has been in existence more than 30 million years, making it one of the oldest mammals on Earth. It also boasts the longest migration of any mammal: an average 20,000-kilometer round trip from the cold Alaskan waters to the warmth of Baja California, Mexico. Once, a number of gray whale populations existed: possibly two in the North Atlantic (now extinct), the western North Pacific (nearing extinction), and the eastern North Pacific (the largest surviving population). The pattern of migration, feeding, mating grounds and calving of the two Pacific populations was different from the other whale populations. The eastern North Pacific gray whale migrates along the Pacific Coast and is frequently visible from the shore as it makes its amazing trek. (Rice and Wolman 1971)

The gray whale has many distinct features that set it apart from other large sea creatures. Its unmistakable profile includes a narrow, tapered head, the lack of a dorsal fin, and a sequence of bumps (often ten) along the middle of the lower back. Individual whales can be identified by multiple scratches on the skin, scattered patches of white barnacles, and orange whale lice. When gray

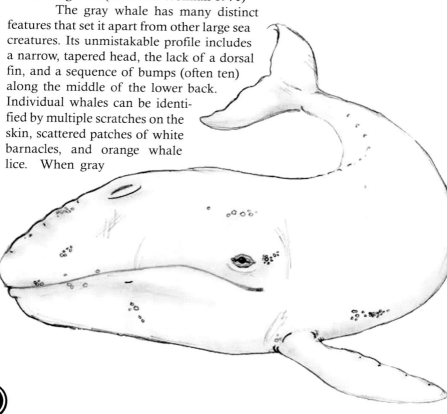

whales begin their northern journey from Baja California, females whales frequently have newborn calves in tow, distinguished by dark gray or black coloring and white markings. (Watson 1981) (Rice and Wolman 1971)

The gray whale has a curious nature. On occasion, a lucky ocean visitor has the extraordinary experience of encountering "friendlies" that approach boats and do not mind being touched by humans. The Baja California breeding and calving area is where this unique event is most often experienced.

The gray whale breeds from winter to early spring. It reaches sexual maturity between five and 11 years of age. Courtship and mating behaviors include spy hopping (a sudden thrust of the head straight up out of the water) and circling. The

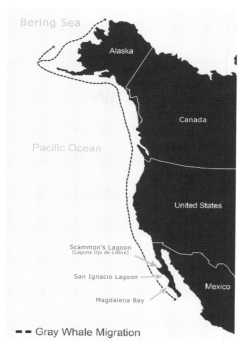

- - Gray Whale Migration

Migration route of the gray whale. It has the longest migration of any marine mammal. The round trip is over 20,000 kilometers.

gestation period lasts from 12 to 13 months. The female gray bears a single calf (about about the size of a Volkswagen Beetle) at intervals of two or more years. The calf remains with its mother until it is weaned—usually in the summer following its birth.

A full-grown gray can measure more than 15 meters and weigh more than 27,000 kilograms. Since the whale is so large, it needs an enormous amount of nourishment to survive—more than 60,000 kilograms of food annually, which is about the same amount of food the average human consumes in a lifetime. The gray dredges the ocean bottom for food such as amphipods, polychaete worms and mollusks using its baleen plates (fringed plates that hang from the upper jaw) to filter out mud. When it closes its mouth, water and mud rush out through these plates and trap food. Interestingly enough, when feeding, the gray whale will invariably roll on its right side to suck up substrate spitting a large cloud of filtered dirt from the upper, left side of the mouth. ("Gray Whale" 2007) (Sidenstecker 2007)

It is difficult to tell the age of a gray whale since it has no teeth. However, its estimated lifespan is 50 to 60 years. Because of its immense size, the gray has few natural predators, although occasionally it falls prey to an orca (killer whale) attack. Pods of orcas tend to attack the smaller or younger grays as they are easier prey.

The gray whale enjoys frolicking in the water and sometimes appears to surf the waves. Surprisingly, it is an agile swimmer in spite of its immense size. The gray whale can hold its breath for as long as 30 minutes and can dive to a depth of 150 meters. When one thinks of the gray whale, the most common image that comes to mind is a *National Geographic* shot of one defying gravity as it leaps in the air. This act called breaching, which is accompanied by a dramatic spray of water and is believed to help clean off barnacles and whale lice. It may also be a form of communication between whales, and provides a magnificent spectacle for enthusiastic whale watchers. (Gordon and Baldridge 1991)

Prior to its overexploitation the gray whale migrated in great numbers off San Diego's coast. Richard Henry Dana described the abundance of this species during his voyages along the California Coast between 1834 and 1836. "We were surrounded far and near by shoals of sluggish whales and grampuses [Risso's dolphins], which the fog prevented our seeing rising slowly to the surface or perhaps lying out at length heaving out those peculiar lazy deep and long drawn breathings which give such an impression of supineness and strength." (Dana 1840)

Today, patient observers on the shores of San Diego Bay can spot grays. During the course of this study migrating whales were seen from both ship and shore. We were even fortunate enough to capture on film an adult breaching at the mouth of the bay on its return to northern waters. Mid-January is the peak of its migration south, but the "whale watching" season is traditionally mid-December through mid-March. Typically, the eastern North Pacific gray whale swims south without feeding and can be observed at one-minute intervals as it surfaces to breathe. It is more common to encounter a gray feeding during its northern route to Alaska.

Errant gray whale that wandered into San Diego Bay, March 2009.

O, GRAY WHALE

O, gray whale,
Barnacles line your wry face.
You are tired, so tired
From swimming from place to place.
Breaching along the rugged coast
Watching as others of your kind are taken,
Living in a world where your presence is forsaken.
Such a thing is too horrid to be a lie.
The wake of a ship crosses your path
Flee from the dreaded man
Go with all of your might
Flee as fast as you can.

— Jessica Shutt

DECLINE OF THE
Gray Whale

Notations in the personal diaries of early 1800s traders commented on the existence of whales in San Diego Bay. However, whaling in the area was nonexistent thanks to restrictions set down by the Mexican government. In addition, the relatively combative nature of the gray whale gave it a distinct advantage when facing off against the hand-held spears commonly used by whalers during that time, so it was safe from humans. However, in mid-nineteenth century a type of harpoon gun, which shot explosive bullets, was developed. Moreover, San Diego had then come under the control of the United States, and some miners from the California Gold Rush, realizing that they were unlikely to strike it rich, looked for occupations in other industries—including whaling. Whalers sailed into the lagoons along the Baja California coast and took females in great numbers. ("Gray Whale Migration Route" 2008) (May 2007)

In 1857, the brothers Prince William and Alpheus Packard arrived in San Diego. They based their shore-whaling operation in La Playa and established a plant on Ballast Point that rendered whale flesh into oil. The brothers, and several other whaling camps, met with great success. Ballast Point was an ideal place because the whales, being coastal migrators, came in close to shore. It also kept the whale-oil plants well away from San Diego residents as they were "horribly smelly—the burning oil was unbearable for most people." (May 2007)

In 1869, Isadore Matthias invested in the Johnson and Tilton and the Packard whaling companies. The Ballast Point whaling stations and those just down the coast at Punta Banda and Santo Tomás harvested more than 20,000 gallons of oil in the 1870–1871 season. The next season yielded 55,000 gallons, and by 1873 the gray whale population had dropped significantly. Consequently, it was probably fortuitous that in 1873 the whalers were evicted from Ballast Point by the U.S. Army, an event that marked the end of their financial backing by Matthias. Various whaling companies from San Diego continued to hunt grays until 1886, but not nearly to the extent of the early 1870s and by 1887, whaling had died out as an industry in San Diego. The whalers drove the gray whale to near extinction; had they continued whaling after 1887, it is possible that there would be no grays today. (May 2007)

The dive of the gray whale is not as deep or as long as that of other whales. Consequently, it was an easy target when migrating through

narrow coastal corridors for its annual, traditional return to the shallow calving lagoons in Baja California. In 1947 the International Whaling Commission agreed that the gray whale should have full protection status. Unfortunately, this did not effectively stop hunting. ("Gray Whale Tutorial" 2008) (Gregr 2000)

Global warming is believed to have caused a fluctuation in gray whale calf production. The Southwest Fisheries Science Center has reported that calf births from 1994 to 2000 fluctuated because of changes in the seasonal ice cover at the North Pole. Moreover, scientists fear that the increasingly frequent sight of emaciated gray whales along the Pacific Coast is an indication that "global warming is wreaking havoc in the whales' Bering Sea summer feeding grounds." ("Save the Whales" 2007)

The population of amphipods that once constituted the whale's primary food source is declining, and plankton is being killed by increasing amounts of chlorofluorocarbons as greater ultraviolet radiation passes through our atmosphere. Because the gray needs more than 60,000 kilograms of food per year it is being forced to seek out a much more diverse prey base. Even though humans may not be killing the whale directly, every time they drive automobiles and send more carbon dioxide into the atmosphere, they may be contributing to the degradation of resources the whale needs to survive. (Perryman, et al. 2002) (Sidenstecker 2007) ("Gray Whale Tutorial" 2008) (Gregr, et al. 2000)

RECOVERY OF THE
Gray Whale

In 1973 the Marine Mammal Protection Act (MMPA) was established by the U.S. government. The act acknowledged the endangerment of a variety of marine mammals and placed restrictions on their exploitation. Recent Marine Mammal Stock Assessment Reports, compiled and produced by NOAA, have shown a dramatic increase in the gray whale population. In 1968 there were only about 13,000 grays, but in 1994, with a population of about 24,000, the eastern Northern Pacific gray whale was removed from the Endangered Species List because it was no longer considered threatened or endangered under the Endangered Species Act. However, the whale's population fell significantly from its 1998 peak of 28,500 to about 18,500 in 2001—a drop of more than 10,000 whales in only four years. ("Gray Whale (*Eschrichtius robustus*)" 2007)

The relative success of the rehabilitation of the gray whale is partly credited to the drop in the animal's population that took place during World War II when the expansion of industrial whaling prompted a depletion of most of the great whale species. As a result, there were more dramatic increases in the gray population than usual, possibly because its primary predator, the orca, had declined in numbers.

In California, military warfare training exercises, using high-powered sonar devices have drawn skepticism from environmental groups as they pose danger to the gray whale and other marine mammals. As recently as the summer of 2006, a California district judge placed a restraining order on one such training exercise, prompting the U.S. Department of Defense to give the Navy a six-month exemption from the MMPA. In January 2008, the same judge ruled that the Navy could not deploy high-powered sonar devices within 12 miles of the California coast. In an unprecedented case, Save the Whales prevented the occurrence of 270 "Ship Shock" underwater Naval tests from taking place over the

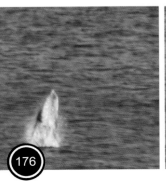  

course of five years in the Channel Island Marine Sanctuary. Perhaps as many as 10,000 marine animals would have been killed outright as a result of such tests. (McClure 2008) ("Save the Whales" 2007) (Sidenstecker 2007) ("U.S. Navy in Sonar Ban over Whales" 2006)

Organizations like Save the Whales, which provide insights into the negative human impacts on whale populations, are currently the best means of educating the public and raising general awareness of the fragile position of marine mammals. Ron May, a historian who studied the nineteenth-century Ballast Point whaling camps believes that education such as that provided by Save the Whales has played a vital role in the gray's recovery: "the preservation of their habitats, [and increasing] world awareness of the plight of the whales, that's really why they are back. [However] we always need to be vigilant that somebody won't come back in someday and find an economic use to kill the whales that overrides the need to preserve them." (May 2007) ("Save the Whales" 2007)

Maris Sidenstecker, founder of Save the Whales, when asked which accomplishment she was most proud of in her conservation work stated, "two things: educating over 277,000 students with our hands-on outreach program 'Whales On Wheels'™ and stopping the Navy from "Ship Shock" testing in a Marine Sanctuary." (Sidenstecker 2007) During migration season, Save the Whales' volunteers go out on boats and untangle unfortunate animals that have been caught in stray fishing lines. Save the Whales prides itself on its investments for the future, realizing that there is a great deal of under-utilized interest in whale population preservation, so they feel that it is their job to rally a call for action. ("Save the Whales" 2007) (Sidenstecker 2007)

In addition to Save the Whales, other organizations such as Greenpeace and Sea Shepherd have played considerable roles in the conservation of gray whale populations. Both organizations contribute to the well being of grays, although each organization goes about this in a different way. Greenpeace believes in nonviolent protests against organizations that harm gray whale habitats, while Sea Shepherd takes a more aggressive approach to those who interfere with the whale. Greenpeace, established in 1971, has been a presence in oceans all over

the globe to prevent whale poaching of all kinds as an attempt to rally support for the next convening of the International Whaling Commission. Sea Shepherd has attracted a great deal of criticism because of their unorthodox approaches and destructive protest methods.

The gray whale is one of the few of the great whales to have an infant raised in captivity. On January 10, 1997 near Marina del Rey, California a malnourished and stranded infant gray, no more than one week old, was found in knee-deep water. At the mercy of the receding tide, the infant struggled to stay afloat. When it was discovered that there were no migrating grays in the area, it was decided that to save the whale's life, the female would have to be transported to SeaWorld San Diego without fail. The trip itself took approximately three hours, with veterinarians continually pouring water on the infant to keep it comfortable. (Heyning and Heyning 2001)

The night of the whale's arrival was filled with trials and tribulations for the young calf. Since the infant could not stay afloat on its own, it was placed in a shallow tank, which did not require it to sustain buoyancy. Throughout the night, rescue workers rehydrated the calf and gave it supplementary nutrients and antibiotics to strengthen its weakened immune system. SeaWorld veterinarians then spent the remainder of the night developing a formula for the infant— based on other whales that had been successfully rehabilitated by their program. (Heyning and Heyning 2001)

After two days of work, the veterinarians started the infant on the synthetic formula. At first, feeding was quite difficult, but it learned quickly, and soon associated humans with food. The day after being fed two gallons of the formula the infant gray gained 14 kilograms and would go on to gain around 36 kilograms during its first five days at SeaWorld. When the SeaWorld veterinarians noticed significant improvements in the calf, they decided on the name J.J. to commemorate the late Judi Jones, a longtime advocate of the San Diego sea lion population. Though J.J.'s astounding initial growth rate began to level off, it gained a total of 181 kilograms during the first two weeks at SeaWorld. (Heyning and Heyning 2001)

A month into its stay, J.J.'s weight reached 408 kilograms, so the animal was moved to a larger tank. After eight months it finally learned how to feed at the tank's bottom, thus enabling it to survive in the ocean. The young gray was released on March 23, 1998. J.J.'s rehabilitation is widely considered to be a success, though its whereabouts are unknown. (Heyning and Heyning 2001)

Much was learned from observing the gray whale calf. Here, J.J. feeds on squid at the bottom of the SeaWorld tank.

The successful recovery of the gray whale population gives one hope for the future of marine mammal conservation. Of all the whale species, the gray is the only one to have met with such success. The recovery plans must continue and hopefully more can be learned from this story so that other species that are currently on the brink of extinction may be saved.

Maris Sidenstecker

Devotee to conservation efforts, Maris Sidenstecker founded Save the Whales when she was 14 to promote education about marine mammals and the oceanic environment. This internationally recognized organization has blossomed since then, and is responsible for educating more than 277,000 children to the plight of whales through the program Whales on Wheels™. Sidenstecker's desk is littered with paperwork for her various projects, such as saving countless marine mammals from deaths caused by Navy "Ship Shock" tests, and for preventing salt mining in the last remaining undeveloped gray whale birthing bay. The papers on her desk are merely symbolic of her continuing efforts to protect marine mammals from the hardships they face from mankind's actions.

Student Researcher (SR): How large does the eastern North Pacific gray whale population need to be so it is no longer a concern?

Maris Sidenstecker: This is hard to estimate. The population of 20,000 plus animals is still not a very large population. An overlooked potential danger is the lack of genetic diversity within this species and absence of genetic studies. The effects of global warming and chemical pollution on future generations pose serious questions that lack answers.

SR: At what point did the nation recognize the plight of the gray whale and other great whales?

Sidenstecker: The great whales have been protected since 1986 when there was an international ban on commercial whaling. Even with protection, the only great whale to make a significant recovery is the gray. Due to loopholes in the International Whaling Commission regulations, Japan, Norway and Iceland have been whaling on the small baleen whales or minke whales.

SR: What was your inspiration behind Save the Whales?

Sidenstecker: When I was 14, I went to Boston to visit my father during a summer vacation. The airline magazine contained a story on whales and how cruelly they were killed. I was shocked to read about a pregnant blue whale that had suffered on a dock for several days before dying. I felt people had to know what was happening to the whales. I told my father, who was an art director, that I wanted to design a t-shirt for the whales. He encouraged me to work on the design and what I wanted the message to be. After much thought I decided what I most wanted was "Save The Whales." This became the message on the shirt along with my drawing of the blue whale.

SR: What are you most proud of as a marine advocate?
Sidenstecker: There are two things. First of all, we focus mainly on education and developed an innovative hands-on program called Whales On Wheels™ (WOW), which we brought to hundreds of thousands of children all over California both in English and Spanish. WOW has been fortunate enough to also travel all over the country.

The biggest effort of Save The Whales was our battle to stop the Navy from performing "Ship Shock" tests in the Channel Islands Marine Sanctuary, a biologically sensitive area off the coast of Southern California. These waters are home to blue, sperm, fin and humpback whales, as well as other marine animals. If the U.S. Navy went ahead with its plans—which was to test the hull integrity of its new cruisers by detonating 270 underwater explosives—it was estimated that 10,000 marine mammals, including many endangered whales would have been killed outright.

We took this issue to court in the U.S. District Court, (Central District), in downtown Los Angeles. Judge Stephen V. Wilson presided and at the end of the five-day hearing, he found that "the Navy had failed in its obligation to protect marine mammals; that it hadn't prepared a full environmental impact statement; and that it hadn't investigated all reasonable alternative sites and properly mitigated the impact of detonations on marine life." In the end, one detonation would be allowed farther offshore with observers of our choice, including airplanes, and instruments would be used to detect any deep-diving marine mammals.

SR: What do you believe needs to change for the increase of global whale populations?
Sidenstecker: The following: (1) a new governmental leadership that will have an environmental policy; (2) the cessation of low-frequency underwater weapons testing, which is found to damage cetacean hearing and has been linked to beached whales; (3) protection of ocean habitats; (4) preventing urban runoff; (5) curtailing global warming that affects plankton, the beginning of the entire ocean food chain; (6) stopping whaling completely.

California Sea Otter

(Enhydra lutris)

The sea otter is a highly intelligent marine mammal which uses crude tools such as small rocks to break open clams, abalone, and purple sea urchins that are part of its diet. A member of the weasel family *Mustelidae*, it sports as many as eight million hairs on its body—the thickest fur in the animal kingdom. Upper guard hairs and a thick undercoat trap air, keeping water from touching the otter's skin and allowing it to stay warm despite its lack of insulating fat. Sea otters have other unique features including phenomenally strong jaws and a pouch of skin underneath each forepaw for storing food collected during dives. The front paws also have retractile claws, while the hind ones are longer, flattened and webbed. Its tail is short, thick and muscular. A full-grown female can weigh as much as twenty-eight kilograms and the male can weigh up to forty kilograms. ("Sea Otter" 1998)

This sea otter is enjoying a large rock crab, eating its catch in typical otter fashion, while floating on its back.

In California, the majority of sea otter pups are born between January and March, although the female otter is able to reproduce at any time of the year. The pup normally weighs anywhere from one to three kilograms at birth. Young otters cannot dive because of the buoyant force created by the air trapped in their fur, and must therefore rely on their mothers for the first eight months of life. The body and muscle mass of a pup is far less than that of an adolescent or adult otter. ("Sea Otter MMC" 2002)

The sea otter is a social animal with females and pups congregating in one group and males in another. When the otter is at rest, it usually lies on its back with paws crossed, similar to the pose of a body lying in a coffin. It is sometimes seen covering its eyes so it can sleep in the sunlight. The sea otter can be found in shallow waters near kelp beds, where it uses the kelp to anchor itself and keep from drifting out to sea while resting. ("Behavior of the Sea Otter" 2006)

The sea otter is an excellent example of a keystone species—a species that plays a critical role in maintaining the balance of its respective ecosystem. The sea otter aids kelp forest habitats by helping to regulate abalone and sea urchin populations that, without the mammal's help, could excessively feed upon kelp and algae with devastating results. A significant decline in a sea otter population can have a huge effect on kelp forests, resulting in an unhealthy environment for other native organisms. Sea otter conservation is thus understandably vital to the preservation of California's coastal ecosystems. ("See Otter MMC" 2002)

DECLINE OF THE
Sea Otter

This commemorative coin depicts the *Lelia Byrd* conflict in San Diego Bay over smuggled sea otter pelts.

As a keystone species, the decline of the sea otter was likely disastrous to the stability of San Diego Bay's delicate biological balance. The lucrative nature of the fur trade sealed the demise of the sea otter. It is believed it was the fur trade, more than any single event or resource, that opened up the California coast to world trade. (Skinner 1962)

The real decline of the sea otter originated in the 1740s when Russians and Europeans began hunting the animal for its thick, highly prized fur. Russia's economy in particular began to depend on the fur trade, and an increasing number of hunters embarked on short voyages to find the animal. After the otter population in eastern Russia's Commander Islands became severely depleted, these voyages became longer. Russian hunters discovered a large sea otter population in the Aleutian Islands off Western Alaska and took hostage Aleut hunters to help capture the animal. Once the Aleutian otter population had also been depleted, the Russians ventured closer to the American mainland, often as far as the Northern California coast, on voyages that could last as long as three to five years. (Estes 2002) As early as 1804, Russians brought nearly 100 Aleuts to the coast of California to hunt sea otters from their canoes made of animal skin. (Skinner 1962)

Americans also were involved in the fur commerce, trading some sea otter pelts to China. American hunters were astonished to discover how much the Chinese would pay for the pelts and this fueled a fierce competition with the Russian and European traders, even leading to a battle in San Diego Bay involving a pelt smuggling brig, the *Lelia Byrd* in 1803. This battle was the beginning of an international dispute between Russia and the United States.

The value of the fur of this unique marine mammal and the profits made from their exploitation nearly parallel the Gold Rush that would soon follow in the foothills to the east. In 1784 a Spanish expedition traveled the coasts of California and Mexico to procure sea otter fur. They exchanged various trade goods, including beads and metal articles for sea otter pelts from the local Indians of Northern Mexico and Southern California. It was reported by a Spanish expedition leader, Senor Vincente Vasadre y Vega that from 1786 to 1790, nearly 10,000 sea otter pelts and an unknown number of seal skins were sent to Manila, which brought an amazing sum of $3,120,000. (Skinner 1962)

In 1812, the the Russian American Company became established in Central California and became a leader in sea otter pelt harvesting. Worried by the dwindling otter catch in Alaskan waters, this new Russian-led venture had dispatched scouting expeditions to California in 1803 in collaboration with American sea captain, Joseph O'Cain. The explorative hunting party sailed as far south as San Diego Bay and Baja California in Mexico. The scouting voyage found sea otters to be in abundance in this area, which ensured that the pelts would remain the company's most lucrative trade item. A hunting and processing station was also established with Aleutian Indian hunters on the Farallon Islands and the following years of hunting in California waters were quite successful for the Russians. This overexploitation eventually resulted in the California sea otter population becoming decimated by the mid-1830s, and the Russians shifted their emphasis from hunting to farming and raising livestock. Under Mexican rule, the hunting of sea otters became illegal in California. However, smuggling of sea otter pelts was rampant and hunting of sea otters and sea lions was practiced illegally by several nations.

By 1911, sea otter populations in the Pacific had diminished so significantly that an international treaty was reached to prevent the hunting of the marine mammals. (Estes 2002). The depletion in the California sea otter population that occurred in just over a century is nearly unparalleled. A manuscript by General Mariano Vallejo referred to the abundance of the sea otter in San Francisco Bay: "They were so abundant in 1812 that they were killed by boatmen with their oars in passing through the kelp." The sea otter had played a role in the settlement of California, but they were nearly driven to the brink of extinction until a vastly overdue state law in 1913 protected the remnant population along the coast of San Simeon, south of Monterey Bay. (Skinner 1962)

RECOVERY OF THE
Sea Otter

After the 1911 treaty and the state law of 1913, California's sea otter population steadily rose, with an average growth rate of 5.5% per year. Starting in the mid-1970s however, nearshore gill-net fishery caused a dramatic decrease in its numbers so the nets were moved farther offshore. Since then, California's sea otter population has experienced occasional fluctuations but has remained fairly stable. A 2004 estimate was 2,500 individuals, roughly 15% of the believed pristine state. Though the California sea otter is still listed as threatened under the Endangered Species Act, its recovery is considered to be one of the great accomplishments in marine conservation. (Estes 2004[1])

Although sea otter hunting is no longer commonly practiced, the marine mammal still faces a number of threats today. Perhaps the most serious of these is the risk of infection by a number of parasites, especially *Toxoplasma gondii*. This protozoan parasite originates primarily in the digestive systems of cats, and is released into the environment by the cats'

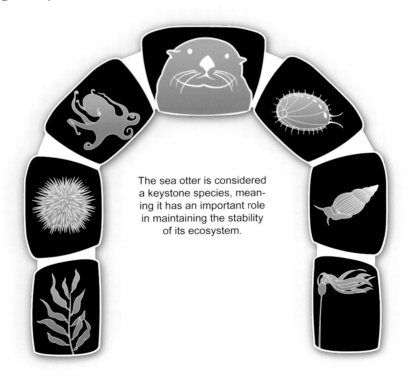

The sea otter is considered a keystone species, meaning it has an important role in maintaining the stability of its ecosystem.

feces. These droppings can make their way to the ocean by surface runoff, storm drains and sewage spills, eventually arriving in otter habitats. The parasite may then infect the sea otter and cause toxoplasmosis, a disease that has been known to cause fatal brain lesions. Dr. Patricia Conrad of the University of California, Davis found in her California Sea Grant project that 42% of otters carried an antibody to the parasite and were probably infected. If the disease were to spread on a large scale, the sea otter could face a serious crisis. (Conrad, et al. 2005)

A number of experts, including renowned sea otter specialist, Dr. James Estes of University of California, Santa Cruz, believe that orca whale predation may be responsible for decreased otter numbers in recent years. Reduction in habitat and natural food sources for the whales may have caused them to prey upon the otter and other marine mammals. This theory, while plausible, remains controversial and largely unstudied, relying primarily on circumstantial evidence. (Estes 2004[2])

A notable ally of the sea otter is the Monterey-based organization, Friends of the Sea Otter. This group contributes to the animal's recovery efforts by gathering population data to determine if the California sea otter is indeed making a comeback. Population monitoring is especially important because, although sea otter populations are today healthy, a sudden, unattended downward spike could lead to the extinction of the species. If the population is monitored properly, such an event might be avoided or, in the very least, become apparent before it is too late to act.

(Below left) Only captive otters at SeaWorld remain in San Diego today. Wild otters along the Central California coast have come back from the brink of extinction.

STORY OF THE
SWEETWATER RAINBOW TROUT

The surviving male Sweetwater rainbow trout.

High Tech High juniors participate in an internship program to get experience in a professional environment; I was fortunate enough to intern at the Chula Vista Nature Center (CVNC). When I arrived for my second day of work in January 2007, I was immediately informed that something important had happened the previous day. Earlier that month, CVNC had been able to fertilize the eggs from the world's last remaining female of a local rainbow trout subspecies (*Oncorhynchus mykiss*) using the sperm of the last remaining male. While the fertilization appeared to have been successful, the female had died overnight and the male's health was rapidly declining. It was apparent that immediate action was necessary.

This population of rainbow trout, native to the Sweetwater River watershed, was nearly eradicated in the past few years. The tragic and highly publicized 2003 wildfires that destroyed thousands of acres of San Diego County greatly impacted the headwaters of the Sweetwater River ecosystem, contaminating the watershed with ash and even drying up entire portions of the river. The struggling Sweetwater rainbow trout was further devastated by a flash flood in December 2006. Though CVNC already had several members of this endangered subspecies as part of an onsite exhibit, successful captive breeding was limited and many of the original fish had died from old age and natural causes. A 2006 survey of the Sweetwater River revealed that the center's remaining male and female were probably the last of the subspecies. (Rodgers 2007)

The mood was solemn as the center staff discussed what needed to be done to save the sickly male. Dr. Brian Joseph, the CVNC veterinarian, thought that a procedure to remove harmful bile was necessary to save the fish's life and that increasing the salinity in the water could benefit its health. With this treatment plan in order, I was tasked with adding saltwater to the male trout's exhibit. About 150 gallons of freshwater were siphoned from the holding tank, and this volume was replaced with saltwater to give a 10%-saline solution. By the end of the day, the assignment had been successfully completed and, despite the trout's initial poor prognosis, its health began to improve.

Unfortunately, a few weeks later the eggs from the expired female were found to be infertile. Cells from the eggs and body were sent to the Conservation and Research for Endangered Species Center at the San Diego Wild Animal Park where they were cultured and placed in a "frozen zoo," an archive of viable cells and DNA. Fortunately, the outcome for the male was much better; he recovered completely and continues to live at the CVNC. This story was considered so ecologically important that it was featured in the metro news section of the *San Diego Union-Tribune* in March 2007.

As to the future of the Sweetwater rainbow trout, there is some interest in crossing the remaining male with a female from a related strain. By doing this, it is hoped to create a hybrid that could potentially fill the original subspecies's role in the Sweetwater River ecosystem. However, others disagree with this course of action, echoing a division among conservationists about the principles and methods of conservation as a whole. Dr. Joseph explained that aggressive efforts to preserve individual species, such as the California condor, the giant panda and the local Sweetwater rainbow trout were viewed by many pseudo-conservation. These individuals believe that conservation should not, in the words of philosopher Henry David Thoreau, be an effort that "hacks at the branches of evil," but instead one that "strikes at the root" of ecological problems. In most cases, it is not the efforts involving well-known, threatened or endangered animals that truly make the difference. It is usually more important to address major problems such as habitat destruction, over-development, pollution and endangered species protection as a whole, to prevent them from becoming threatened or endangered in the first place. Perhaps the best solution is a balance between the two philosophies; while it is important to protect and rehabilitate individual species, more effort should focus on the bigger picture.

The important difference between the Sweetwater rainbow trout and species such as the light-footed clapper rail and others described in this book, is that hope remains for the rail. Reflecting on the situation of the trout, CVNC aquarist John Lopez said in the *Union-Tribune* article, "The world just lost a small part of its biodiversity. What strikes me is this: there's no ceremony. No event is held to acknowledge it. It's a quiet thing." It is vital that the world learn from unfortunate cases such as the Sweetwater rainbow trout and work actively to preserve natural ecosystems and endangered species.

REASONS FOR DECLINE

"The one process now going on that will take millions of years to correct is the loss of genetic and species diversity by the destruction of natural habitats."

— *Edward O. Wilson*

Ship Traffic
Invasive Species
Dredging
Land Use
Climate Change

Ship Traffic

"… the Port of San Diego, more than anyone in the country right now, is pushing the development of alternative boat-hull paints that don't have toxic materials in them."

— David Merk

Ships have long sought harbor along the rocky and rolling cliffs of the California Coast. Tucked behind Point Loma, San Diego Bay provided mariners with relief from the open sea in the form of narrow channels meandering through vast marshlands.

Since Juan Rodriguez Cabrillo landed in his flagship, the *San Salvador*, on September 28, 1542, sea vessels of all types have found refuge in the gentle harbor that is San Diego Bay. Though Cabrillo's intention was to find a route from Spain to China, he recognized the significance of his discovery and gave the bay its original name, San Miguel Bay. ("San Diego History Timeline" 2007)

San Miguel Bay underwent two periods of ownership before becoming part of the United States. The first, the Spanish period, began in 1602 when explorer Sebastián Viscaíno arrived. Viscaíno was directed

Few events in the history of San Diego Bay, and the entire Western Coast of North America, have been as significant as the landing of Juan Rodriguez Cabrillo. Above is a reenactment of the event.

not to change the name of San Miguel Bay, but nonetheless renamed it after his flagship, the *San Diego* and the fifteenth-century saint. Spanish presence in San Diego was strengthened with the arrival of Don Gaspar de Portolá a century-and-a-half later, in 1765. Portolá established a military fort in 1769—the Royal Presidio of San Diego—that still stands today and overlooks Old Town. The Presidio acted as a magnet for San Diego Bay, attracting a multitude of ships in the following years. ("San Diego History Timeline" 2007)

After the War of Mexican Independence ended in 1820, San Diego Bay no longer belonged to the Spaniards and was transferred to Mexico. This period, however, was short lived. Much of what we know about San Diego Bay's beginnings as a center of commerce comes from the writings of New Englander Richard Henry Dana, who sailed aboard the *Pilgrim* in 1834. In *Two Years Before the Mast*, his account of the voyage, Dana described his experiences in San Diego Bay and along much of California's coast. (Dana 1840)

During the Mexican-American War, the officers and crew of the *U.S.S. Cyane* became the first to raise the American flag on San Diego's shore. With the California Gold Rush, San Diego's population started to rise dramatically in 1849, requiring both shoreline safety

and residential development. When California became a part of the United States a year later in 1850, the government agreed to build lighthouses on its coast. The first of these was constructed in Point Loma in 1855. ("San Diego History Timeline" 2007)

Five years later, San Diego Bay was filled with ships from all around the world. Aboard some of these vessels were individuals who would become significantly involved in creating today's City of San Diego—including Alonzo Horton, the founder of "New Town," (now downtown San Diego). New Town became a huge center for commerce; its bayside location brought an increasing amount of ship traffic to the area and the waterfront bustled with activity. Both sailing and steam ships frequently entered San Diego Bay to deliver building materials, coal, food and those who wanted to live in the area. San Diego Bay's quick

rise in popularity was a direct result of its unique location; it is the United States' last port-of-call for ships sailing south, and the first port for ships sailing north. ("San Diego History Timeline" 2007)

Today, because of its warm weather, San Diego is considered to be one of the best and most desirable destinations for year-round travel. The bay has many recreational opportunities, and as such attracts a large number of tourists annually.

The 57 square miles of the bay serve as home port for a commercial fishing fleet, the 11th U.S. Naval District's headquarters,

The San Diego Maritime Museum preserves historical vessels such as the tall ship *California* shown here.

Much of the contemporary ship traffic in San Diego Bay is dominated by the U.S. Navy. The bay is home to 25% of the Navy's Pacific Fleet.

an operations base for the U.S. Army, Marine Corps, Coast Guard forces, and innumerable recreational boats and yachts. In addition, San Diego Bay serves as a commercial port for both domestic and international shipping. Imperial Valley's cotton industry is one of the many industries that ships products from San Diego Bay.

The bay's traffic consists of about 10% military ships, 50% commercial vessels and 40% recreational boats. There are tourist and commercial areas on both Harbor and Shelter Islands, and marine industries, such as National Steel and Shipbuilding Company (NASSCO), are located near the San Diego-Coronado Bay Bridge. All of this contributes greatly to San Diego's economy. Despite its current development, the City of San Diego is still capable of further expansion. (Merk 2007)

Although areas such as San Diego Bay greatly contribute to the economies of nearby cities and allow commercial and military operations to run much more smoothly, they also become reservoirs for pollution in the form of toxic runoff and trash spills. Bays also become further polluted by the toxic paint that peels from boats and falls into the water, and sewage from boats that is dumped directly into the water. Proper care and maintenance of boats is essential for preventing such toxins and other contaminants like rust and trash from entering our bay. (Merk 2007) (Chase 1999)

One of the most harmful "pollutants" comes in the form of aquatic invasive species, often introduced by ships traveling through different ecosystems. More information on this concern can be found in the Invasive Species chapter (page 200). ("Stop Aquatic Invaders on Our Coast!" 2006)

San Diego's 1.2 million inhabitants contribute a great amount of pollution to the bay's 122 kilometers of shoreline. Those who participate in recreational activities often are not aware of the effects of their actions, and consequently are not as careful as they might be. Human disturbance of marine animals from boating and recreation is a primary concern of conservationists. The very presence of people and boats disturbs many native species, and when animals are forced out, an imbalance is created to the ecosystem, which can be detrimental to its entire function.

The delicate ecosystem of the bay has been abused and exploited over the years, causing possibly irreversible damage, and it continues to be threatened by further pollution and harm by persistent growth and industrial development. Raising community awareness of the bay's fragile nature is the best way to keep it safe and clean. Public service announcements, seminars and pamphlets are a way to reach boaters and other people who enjoy the natural beauty of the bay, and want to keep it safe. (Merk 2007)

Some anti-fouling products are harmful to the environment and marine life, leaving residues in the bay that are extremely hard to remove from rocks, and from harbor sediments. There are alternative boat-washing methods that are not harmful to the environment, and some are significantly less expensive. (Merk 2007)

The bay is home to a broad range of aquatic and sometimes amphibious vessels.

David Merk

In search of insight into the impact ship traffic has on San Diego Bay, we decided to contact David Merk. He is the Director of the Environmental Services Department at the Port of San Diego, and in this role he deals with the environmental impact of ships in San Diego Bay on a daily basis. The Port controls the majority of all nonmilitary activity in the bay and has more information on this than any other source.

Mr. Merk's department at the Port of San Diego is responsible for protecting San Diego Bay's natural resources, limiting the impacts of urban runoff, providing environmental public outreach, and overseeing dredging and site remediation projects. He has most recently been instrumental in the development of the Port's Clean Air Program, which is the first voluntary program of its kind proposed by a California port to identify and control local port-related sources of air pollution. He joined the Port of San Diego in 1997 after more than 20 years of service in the environmental protection field with both public and private entities.

Student Researcher (SR): How significant is San Diego Bay when compared to other bays?

David Merk: Let me answer in two ways. One way is to say that San Diego Bay is pretty unique. If you asked four people what comes to mind when they think of San Diego Bay: one would say the military because the Pacific Fleet is here; one would say commerce because we get a lot of our local products through the bay; one would say recreation because there are pleasure boats; and one would say its natural resources. So, San Diego Bay is unique in that it's trying to do all of those at the same time. The second answer is in terms of the commerce in San Diego Bay. As compared to ports such as Los Angeles and Long Beach, it's miniscule in terms of the commerce it receives. We're what's called a niche market where we receive certain products, but not others.

SR: How much commerce does San Diego Bay attract?

Merk: The growth in our commerce has been significant over the last five years. In fact, it has grown astronomically. We have three marine terminals. One of those is the Tenth Avenue Marine Terminal, where the bulk of commerce occurs. Then there is the National City Marine Terminal, which is almost entirely restricted to automobiles. And then, finally, here in the north, there's a cruise ship terminal. This year alone, there'll be more than 250 cruise ships that visit our cruise ship terminal.

SR: About how many ships go through San Diego Bay daily, weekly and monthly?

Merk: Serving the U.S. Navy, there are more than 10,000 ship movements annually that go on in the bay. There are also lots of harbor craft, such as tugboats, harbor excursions and ferries, that are moving pretty much throughout the day. And then finally, there are over 8,000 recreational boats in the bay's various marinas.

SR: How much pollution do you believe occurs in San Diego Bay from ship traffic?

Merk: There are small amounts of different things that may come in here, such as oils. But the amount of pollution that is coming from ships is very, very limited. What we are concerned about from ships or [recreational] vessels, is the copper from the boat-hull paints. Copper is a biocide and essentially a pesticide. It's applied to vessels and it leaches out slowly into the surrounding water. Barnacles and other fouling organisms absorb the copper and it kills them, which is the intent of the copper. However, copper that's not absorbed by those organisms is being deposited into the water and sediments of San Diego Bay.

SR: How significant are the effects of copper-based paints on marine life in San Diego Bay?

Merk: There's the Shelter Island Yacht Basin, which is the marina area closest to the mouth of San Diego Bay, and is the most highly sought-after area to moor a boat. In the Shelter Island Yacht Basin alone, copper in the water column exceeds that of normal levels in other parts of the bay. The marinas there have been told that they have to take steps to reduce it.

We also know that when boats are cleaned periodically, divers will come into recreational areas and scrape barnacles off the sides of the boat. The type of cleaning that's used on boats in San Diego Bay now is contributing solid particles of copper into the water, and those particles are damaging marine life in the sediment.

SR: What do you believe is the best method of conserving San Diego Bay and reversing the effects of ship-traffic pollution and recreation pollution?

Merk: I think a couple things could happen here. One that's important is trying to find a way of controlling the copper that's entering San Diego Bay. When it's released into the bay, it can have the same effect on wildlife that are swimming or using the bottom sediments of the bay. Our number one concern is finding a way of controlling that. In fact, the Port of San Diego, more than anyone in the country right now, is pushing the development of alternative boat-hull paints that don't have toxic materials in them.

SR: What environmental programs, related to ship activity, have been most successful in San Diego?

Merk: What's happening in terms of ship traffic is happening at the international level. There are a couple of international associations that are trying to regulate what ships do, and probably the biggest source of pollution from ships is air pollution, surprisingly. They're operating, historically, with very dirty fuels and so their emissions are dirty as well. And there are changes that are occurring both at the international level and at the state and local levels. I think the programs that have been most effective have been those that have been initiated by the shipping companies themselves through these international efforts.

RETURNING THE RHYTHM

The ripples of the bay give movement so natural
It's soothing to one's soul.
But when you have to lift your head that's when you'll see it.
A boat races by and it's going fast
The wake creates waves that hit debris and splash
The sound is shattering,
It slaps the rocks, heaves the pebbles, creates chaos
For a second, the natural rhythm will never return
But it slowly does, peaceful as ever.
It's tired
Returning the rhythm was a fight.
How much longer can the bay fend us off?
When will fatigue overcome its might?
When will we win?
When will we lose?

— Chris Nho

Invasive Species

"It would be hard to say that an invasive species has no affect on the ecosystem, especially if it is surviving."
— Jamie Gonzalez

Invasive or introduced species have long been a problem for ecosystems worldwide. Over the course of history, plants, animals and microbes have accompanied man as he migrated across the globe. These introductions of new species as humans moved about the planet have been both purposeful and accidental. (Diamond 2005)

When species disrupt the ecological balance of their new environment they are considered invasive. The most common reasons for introducing species in ancient times were either for survival or cultural nostalgia. Migrating farmers knew only the plants and animals from their homeland. Therefore, to make themselves feel more comfortable in their new environment, they sometimes brought familiar species with them to make their new home feel more like the old. When the British colonized Australia they brought rabbits and foxes because of nostalgia for the animals of the English countryside. Both these species, particularly rabbits, have caused serious ecological problems in Australia with their destruction of native vegetation. People also brought plants or animals along in case they would not have ample food sources in their new home, or familiar food would not be available. (Diamond 2005)

Today, the problem of invasive species is more common and severe because of another human activity: the widespread movement of boats and commercial vessels. Global trade and transport can introduce species into new environments unintentionally, and even the smallest of species can have a large impact.

According to James T. Carlton, an ecologist who has extensively studied the effects of invasive marine invertebrates in Northern California, there are four primary means by which species have been introduced into the estuaries of San Francisco and the surrounding region. These are hull fouling, ballast water, the commercial oyster industry, and the importation of commercial bait and fresh seafood. (Carlton 2001)

Wooden ships that docked in Central California's bays two centuries ago may have been responsible for the introduction of "fouling" marine organisms from the shores of the Atlantic Ocean. Ship fouling gets its name from the organisms that bore, or foul, holes in a ship's hull or attach themselves to its surface. These organisms can find their way into the ship at one location then, in the case of a transcontinental voyage, be released at one of the vessel's docking points. Within a few days, a foreign and potentially harmful species can arrive thousand of miles from its natural habitat.

During the days of the California Gold Rush, many ships were abandoned in San Francisco Bay, allowing fouling organisms to run rampant. To this day, San Francisco Bay is one of the most heavily invaded estuaries in the world. Some prime examples of invasive marine organisms there are the Atlantic barnacle (*Balanus improvisus*) and the Western Pacific shrimp (*Palaemon macrodactylus*). Later the green crab (*Carcinus maenas*) in the region has created an environmental crisis, nearly wiping out the presence of soft clams on the Northern California coast, and creating a significant disruption in the feeding habits of local shorebirds. (Carlton and Zullo 1969) (Van Heertum 2002)

Antifouling paints have been used in the past for controlling the introduction of invasive species. Many boaters used a copper-based paint to repel species that attached to the unprotected hulls. This caused environmental problems as the copper, a toxic pollutant, was slowly released into the water. Further research has shown the development of copper-tolerant species that not only survive on treated ship hulls, but thrive in copper-polluted waters. Fortunately, copper is not the only method for protecting ship hulls; alternative and effective nontoxic antifouling paints for ship hulls are available. However, the threat of species introduction cannot be expected to disappear entirely and more steps must be taken to ensure the threat does not continue. (Carlton 2001) (Gonzalez 2007)

Another common method of introduction is through ballast water, which is collected and distributed into the ship's internal compartments to balance cargo. This water is dumped before or during docking. The problem is that organisms from the ship's original port are released into its destination's harbor. During the last century, the Chilean beachhopper (*Orchestia chiliensis*) was introduced by lumber ships returning to San Francisco Bay, and in the 1950s ships returning from the Korean War introduced the oriental shrimp (*Palaemon macrodactylus*) in this way.

Recent studies have also shown the importance of monitoring shorter voyages that may be responsible for the spread of invasive species. (Wasson, et al. 2001) A number of the invasive species in Los Angeles and Orange Counties are threatening to the environment. If boaters are not careful, these species could arrive in San Diego Bay in the near future.

A third means of introduction is the commercial shellfish industry. Oysters, such as the *Crassostrea virginica* from the Atlantic coast, may act as vectors for certain organisms, which lodge themselves inside the animals' shells, and escape during shipping and handling. Although stricter measures have been enforced in the aquaculture industry today, many invertebrates can still elude these regulations and cause damage. (Carlton 2001) An example of this is the sabellid polychaete introduced from South Africa that nearly destroyed the California abalone industry. It is thought to have been transported to California on imported broodstock animals. (Culver, Kuris, and Beede 1997) There was great concern that this parasitic worm would be introduced to wild abalone populations.

The final means of introduction outlined by Carlton is commercial bait importation. Worms for West Coast bait-and-tackle shops are packed in algae, which may be discarded in local waters after the shipment arrives. These algae may contain numerous invertebrates with the potential for harm to the local ecosystem. The presence of the Atlantic periwinkle (*Littorina littorea*) in San Francisco Bay can probably be attributed to this action, and the Atlantic quahog (*Mercenaria mercenaria*) has also been found in Humboldt and San Francisco Bays. (Miller 1969)

Possibly the most problematic species in San Diego Bay is the Asian or Japanese mussel (*Musculista senhousia*). This mussel hinders the growth of certain native organisms such as eelgrass and various species of clams. A reduction in eelgrass and clam numbers may lead to a declining fish population that may also directly affect bird populations and, to a lesser degree, humans. From this example, one can see how a species such as the Asian mussel can have a devastating effect on many other species and the ecosystem as a whole. (Carlton 2001)

In June 2000, a potentially devastating invasive marine algae, first discovered in a Carlsbad coastal lagoon, raised a serious alarm in both Southern California and the entire nation. Dubbed the "killer algae," the highly invasive seaweed (*Caulerpa taxifolia*) had nearly destroyed the biodiversity of an area of the Mediterranean Sea.

The Southern California *Caulerpa* Action Team (SCCAT) was established. Because of their fast and effective response, all the patches of *Caulerpa* on the lagoon floor were treated with chlorine held beneath a tarpaulin. This method was successful in killing not only the plant but its subterranean parts. The cost for eradicating the "killer algae" was $7 million. Additionally, the value of the loss of other species in the treated habitat was considerable. (Woodfield 2000)

The algae continues to pose a substantial threat to marine ecosystems in Southern California, particularly to the extensive eelgrass and kelp beds. Banning *Caulerpa* in the aquaria trade, educating the public on its threat to local waters, and continual monitoring will help to reduce the risks of future catastrophes.

Many people do not understand the gravity of the situation and have purposefully released species into local streams or bays. Home aquaria are sometimes dumped into nearby waterways when the owners no longer wish to maintain them and are reluctant to flush the contents down the toilet. The local invasion of *Caulerpa taxifolia* is thought to have originated from such action. Plants and animals from home aquaria should never be dumped outdoors into waterways. See www.habitattitude.net for better alternatives.

There are two primary means for slowing the introduction of invasive species. First is the preventative route: international ship-

An invasive Asian mussel (*Musculista senhousia*) was collected on Grape Street pier, DNA barcoded, and identified by students at High Tech High.

ping regulations that require vessels to exchange their ballast water in mid-ocean. Another is the application of the most effective kind of anti-fouling paint. And, to combat species such as the copper-resistant algae *Ectocarpus siliculosus*, regular hull cleaning paired with the application of alternative, nontoxic bottom paints to reduce invasive species. Taking these precautions will hopefully limit the numbers of invasive species and their transport around the world, although the problem will not be solved entirely. (Carlton 2001)

In addition to preventative measures, well-considered and even "natural" abatement processes will have to be undertaken to reduce the problem of invasive species. For example, *Caulerpa taxifolia* can be defeated with chlorine treatment or with a species of slug that feasts on

A close-up of *Caulerpa taxifolia* growing among eel grass shoots in Agua Hedionda Lagoon in Carlsbad, CA.

Domestic and feral cats are predators of many endangered bird species of the bay.

the seaweed. Many other invasive species, however, have no currently known enemies, because scientists have little knowledge of which species may damage an ecosystem. Analyzing life histories and nutritional requirements, as well as natural predators may help solve the problem. Finding out where an invasive species originated is key, because then it can be tested against other species from the same location. For example, this could be a means for locating an effective predatory species to remove the species in question. Establishing trends in abundance and the origin of an invasive is important to understand the species and detect its presence before its numbers become extensive. One recommendation is DNA barcoding. This method would allow the identification of stages of an organism's life history that would otherwise be difficult to classify. During the course of this High Tech High study, zooplankton were identified using this method. Implementing this testing on a larger scale could allow the immediate identification of a pioneer member of a newly introduced population.

A challenge in creating effective action in dealing with invasives is the difficulty in determining the biogeographic origin. There are many invasive species, such as the tubeworm *Mercierella enigmatica* (found in San Francisco Bay) that was thought to be from France, but has now been found to be of Indonesian origin. Misconceptions like this can slow down the process of abating the damage to an ecosystem caused by an invasive; without a background analysis of a specific species, it is difficult to figure out its dietary needs or its physical weaknesses. For an invasion to be stopped, a dedicated team of researchers is required. Funds need to be allocated and the general public made aware of the problem and its severity. (Hyman 1955)

With increased globalization and importation of goods, the introduction of invasive species will probably increase. Therefore, action needs to be taken to avoid large-scale destruction of our local waters. Since prevention is widely considered to be the best path, researchers, politicians and local citizens should work together to bring about a preventative plan for San Diego Bay.

Stricter regulation of marine vessels is necessary to slow the introduction of invasive invertebrate and algal species, but it is not the only solution. Invasive species may be the result of human leisure activities and are thus difficult to control or stop. The best way to end the problem is to educate and warn people about the hazards of these species, and their effect on ecosystems (see www.protectyourwaters. com).

Hopefully, San Diego Bay will prove to be a good case study for how well preventive measures can work. It is doubtful that measures will ever exist to eradicate the present diverse set of invasive species in and around San Diego Bay. However, effective strategies should continue to be researched and developed to halt future invasions.

The Mediterranean mussel, *Mytilus galloprovincialis*, was DNA barcoded by High Tech High students.

Jamie Gonzalez

As a Program Representative, Jamie had worked for the University of California Cooperative Extension/Sea Grant Extension Program since 2002. She received a bachelor's degree from the University of San Diego in Ocean Studies and a master's in Marine Affairs and Policy from the University of Miami. Most recently, she conducted research on the sustainability of water quality and coastal resources. Some of her other work included the prevention of invasive species on boat hulls as well as nontoxic antifouling strategies for recreational boats. She disseminated research-based information to the government, businesses, scientific representatives, in addition to environmental organizations and vessel owners.

Student Researcher (SR): How would you define invasive species?
Jamie Gonzalez: An invasive species is a species not native to the area where it is found.

SR: Do you think that most people who introduce species know what they are doing, or they are clueless about their actions?
Gonzalez: I think most people are not aware what they are doing because if they knew the effects, then they would probably not be doing it. They are just totally unaware that their actions can have consequences.

SR: What are a few of the most common means of introduction?
Gonzalez: Some are introducing invasive species through ballast water, also through aquaculture and, more recently, we have heard about introducing species on boat bottoms.

SR: Could you tell us a little bit more about ballast water?
Gonzalez: Ballast water is needed for ships to even out their weight. When they pull into port, the ships end up emptying out their ballast water as they unload their cargo. So, all the ballast water from their original port is brought over to their destination, which could be another continent.

SR: What percentage of invasive species end up becoming detrimental to their new ecosystem?
Gonzalez: Every species that is invasive into an ecosystem in which they're not native ends up altering that system in some way; so maybe

Boats are occasionally abandoned in the bay and can become hazards for both other vessels and for wildlife when fouling organisms are released.

all of them. It would be hard to say that an invasive species has no affect on the ecosystem, especially if it is surviving.

SR: What are some of the more infamous invasive species?

Gonzalez: In the United States this would be the zebra mussel. That's because it has caused so much harm in the Great Lakes, clogging up water pipes and overtaking the ecosystem there. I think that's the most famous invasive species and now a relative of the zebra mussel is coming into California—the quagga mussel.

SR: What are some of the most current ways of preventing invasive species?

Gonzalez: This is education—educating boaters, for example, on how to prevent bringing invasive species on the bottom of their boats may greatly deter the spreading of these species to other areas.

SR: Do you believe that most governments are aware of the problem of invasive species but just don't have the resources, or do you think they choose to ignore it?

Gonzalez: In some cases, governments might choose to ignore the problem because many times they don't see the direct impact of invasive species. Then again, funding might be a problem as well. In the United States, a 2007 bill was introduced called the National Aquatic Invasive Species Act. So, they are making it a priority to put funding towards invasive species prevention.

SR: What is your opinion on using other species to combat invasive species that are already established?

Gonzalez: This is really risky, because you're going into that ecosystem and introducing something new. There's already an invasive species

208

there and you don't know really what's going to happen to all the other species there after you introduce a new one. So, even though you're there to help by bringing in this new species, I think its impossible to say what could happen. So I don't totally agree with it.

SR: What are some of the reasons that point to prevention being the best approach?

Gonzalez: Prevention is less expensive than eradication, which can cost millions of dollars. If you prevent the invasive species from getting there in the first place, that's going to be less expensive. You're not going to have to deal with the consequences or impacts of invasive species being in the environment.

SR: Is there anything else you would like to say in conclusion about invasive species?

Gonzalez: Invasive species issues are coming more into the limelight. The public knows more about them now. For example, back when I was in high school, I don't remember talking about invasive species very much, but in college we definitely did. So, I think people are becoming more aware, although there are always those who never get the information. And, I do think governments are making it more of a priority.

Dredging

"If Cabrillo were to sail here today, he would see a very different bay than he did years ago."

— Mitch Perdue

Though many do not realize it, dredging is an important process that has shaped, quite literally, many of the world's major cities. In order to appreciate the changes San Diego Bay and its unique salt marsh and mudflat habitats have endured over time, a thorough understanding of dredging is necessary.

Dredging is the process of removing earth from an area where it is undesirable and the term often refers to dredging occurring in water, usually to deepen waterways to allow ships to pass through otherwise inaccessible areas. Dredging is used to construct waterways and channels, to modify coastlines in order to build ports, harbors and docks, as well as to collect sand and dirt in order to fill in other areas. Such activity has created drastic changes in natural habitats around the world. A decade ago, it was estimated that an annual average of 60 million cubic yards were dredged worldwide and that amount has most likely increased since then. (Meegoda et al. 1997)

While dredging has become increasingly prevalent and pertinent in the past few centuries, it is a practice that has existed for several thousand years. Though there is dispute among historians and engineers as to who exactly began to dredge first, it is clear that various aquatic earth-removal techniques were implemented by a number of civilizations. The ancient Egyptians, for example, dredged parts of the Nile River to benefit their crops and build canals for boat use. Some historians credit the first major canal to Nikau II and Darius I, a project dating to the sixth century BCE that ran from the Nile River to the Red Sea.

Ancient cultures near the Indus River also used an interesting, yet now obsolete, method known as "agitation dredging." This involved stirring up earth from riverbeds and simply allowing it to be carried away by the current. While quite clever and resourceful, such a technique was impractical along coasts where tidal currents did not produce the desired effect. The Aztecs employed similar dredging techniques to build canals at Tenochtitlán, serving as an invaluable means of transportation throughout that marshy city. The ancient Romans perfected a simple "bag and spoon" method that was later used widely throughout Britain and is still in occasional use today on minor dredging projects. The bag-and-spoon dredge was operated by two people and was composed of a spade and a pivoting rope system used to raise and transport the load. (Cooper 1958) (Herbich 1975, 1992)

Modern dredging technologies were developed during the construction of canals and ship channels in sixteenth-century England. One of the primary types used was the bucket dredge, which was improved to meet modern dredging demands. Over the years, it evolved into an effective method that is still widely used today. Bucket dredging involves a series of buckets attached to a device like a conveyer belt and a system of pulleys, rotating to extract and remove earth, which is then collected in an attached containment unit. The original design was powered by hand or mule and then progressed technologically to steam and eventually diesel fuel. (Herbich 1975, 1992) (Cooper 1958)

Many other types of dredges can be found today. Grab dredges, also called clamshell dredges, are designed to excavate and lift earth and move it elsewhere. These dredges are usually mounted on backlines—large cranes that are able to haul the heavy device and its spoils. Backhoe excavation devices are somewhat similar, but are considered to be cruder, as they are usually mounted on barges rather than backlines and therefore simply dig and push sediment away rather than actually hauling it. Backhoe dredges are also sometimes used to level uneven seabeds.

Another primary type of dredge used today is the suction dredge. It is acclaimed for its speed and effectiveness in sucking sediment into large containment units or transporting it through pipelines to another location where it is needed. Some suction dredges are equipped with cutting heads, designed to agitate sediment or tightly packed soil, making sucking easier. The suction dredge is one of the more recent types of dredges to be developed and, with continuous improvements in its hydraulic pumps, it has risen to become the most promising dredging method for the future. (Herbich 1975, 1992)

Differently designed dredging machines are engineered to suit the varied physical conditions of the jobs. Some machines sit on land, some operate from the backs of large ships, some are submersible for deep-ocean dredging and others are amphibious. Most dredges are mobile and can thus accomplish a wide array of tasks. Grab dredges and suction dredges have been used most extensively in the modification of San Diego Bay. The latter of the two has been extremely effective in quickly transporting large amounts of dredged spoils via pipeline from one part of the bay to another.

While dredging is often used to create channels through which ships and boats may pass, it is also employed for creating land for development. This type of geographic modification can be

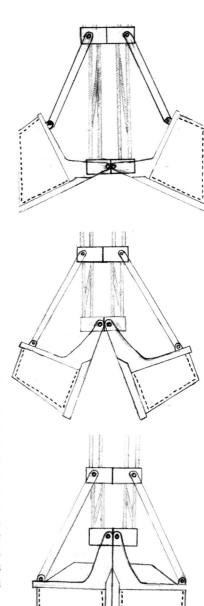

A clamshell dredge, one of the most prominent dredging devices, collects sediment.

seen prominently around San Diego Bay. When comparing recent maps of San Diego Bay with those of the same area from decades earlier, dramatic changes are apparent.

Although dredging has been beneficial to humanity in a number of ways, there are many environmental and ecological implications in altering natural coastlines. Approximately 90–95% of Southern California's coastal wetlands have been completely destroyed, eliminating vital habitat for hundreds of native species. As a result, the populations of many plant and animal species, including the light-footed clapper rail, the California least tern, California cordgrass and other species that previously resided in natural coastal habitats, have declined severely with development. Coastal wetlands, often the victims of dredging plans, are one of the most rapidly declining types of habitat in the world, along with rainforests and coral reefs. (Unitt 2004)

Another important concern in the field of dredging is the disposal of dredged spoils. Soil or sand removed from an area cannot simply be dumped elsewhere haphazardly. Often, human-introduced contaminants are integrated with the coastal sand and soil—especially common at estuaries and other river outlets. Pesticides, toxins and even natural contaminants from upstream are carried to the ocean, where they quickly settle into the sediment. Boat dumping occurs in shipyards where repairs and construction take place and, as a result, pollution levels become especially high in areas with more boat activity. The 2002 discovery of ordnance in San Diego Bay, which originated from ships dumping excess munitions, was so serious that it resulted in the relocation of several environmental monitoring sites for safety reasons. When dredging is performed in contaminated areas, the soil byproducts are often clearly hazardous. Because of this, there are many laws and regulations in place regarding the disposal of dredged material.

There are three general procedures for dredged material disposal; the method used is based on technical, economic and environmental considerations. The first method is open-water disposal, in which dredged spoils are simply disposed of in another aquatic area. This is done with material that either already meets cleanliness standards or has undergone cleansing processes. The second procedure is confined disposal, in which spoils deemed too hazardous are taken to specialized containment areas. The final and arguably most intriguing procedure is when spoils are put to beneficial use—noncontaminated dredging byproducts are used to restore damaged coastal habitats such as marshland. This practice was used to create Shelter Island and North Island in the San Diego Bay region. Like some other small land masses in the area, both of

Dredgers, like this one in the South Bay area, drastically
modified San Diego Bay over past decades.

these were actually sandbars that were transformed into islands by
dredged byproducts. (Meegoda et al. 1997)

Much of San Diego Bay exists in its current form as a
direct result of dredging and subsequent filling. Concerted efforts
to geographically modify San Diego Bay began in the nineteenth
century, when the bay became important for fishermen, travelers
and tradesmen moving imports and exports. Construction in areas
that could not previously be developed and the deepening of ship
passages were some of the first assignments to develop the port. Early
dredging in the bay took place in the downtown San Diego and Old
Town areas to make room for more real-estate opportunities. It was
also necessary for the growing coastal city to have a stable port that
could allow ships—especially the large aircraft carriers of the U.S.
Navy—to pass through without any problems. Between 1949 and
1955, with the onset of the post-World War II industrial mindset,
85% of intertidal habitats in San Diego Bay were lost—an enormous
amount in a very brief period of time. Overall, 92% of San Diego's
intertidal habitats have been lost. (Perdue 2007)

The area formerly known as Dutch Flats in the northern
portion of San Diego Bay was altered extensively by dredging. It was
once a large delta that was the exit point of the San Diego River into
the bay and an estuarine habitat that was likely highly productive at
the peak of its existence. It is now Lindbergh Field, the area on which

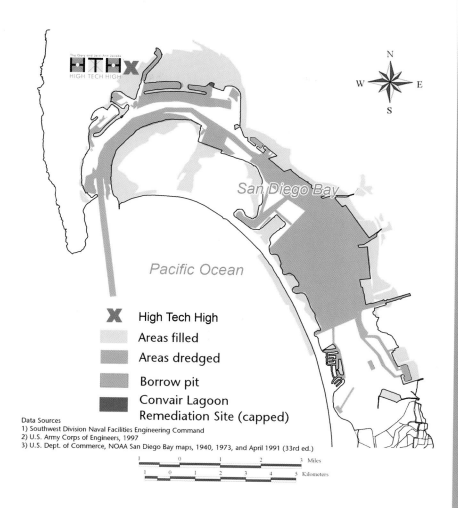

Throughout its history, San Diego Bay has been the site of numerous dredging projects. Areas modified by dredging and filling are highlighted in this map, provided by the San Diego Bay Integrated Natural Resources Management Plan.

the San Diego International Airport is situated. Dredging, modified the land for development, which could never have been accomplished on the previous terrain. Another major dredging project of San Diego Bay that commenced in 2004 is the federal Central Navigation Channel off Coronado and near the San Diego-Coronado Bay Bridge.

Though it is a practice that is typically associated with ecological destruction, dredging has recently been used to help the environment. Dredged spoils and dredging in general have been used to repair damaged

ecosystems and occasionally eliminate environmental problems caused by humans. Perhaps the best example of this in recent years is the restoration of eelgrass sponsored by the Port of San Diego and the U.S. Navy. Since eelgrass supplies refuge and food for many of the fauna of the bay and is generally an important indicator of the health of San Diego's coastal ecosystems, it is important that its population remain stable. Eelgrass can only grow in shallow water because it requires large amounts of sunlight. Some areas around the bay have been made shallower with dredged spoils to provide a habitat for the crucial plant. (Perdue 2007)

One notable restorative dredging project is the construction of Homeport Island, located in San Diego Bay just a few hundred meters from Coronado and the Silver Strand. Homeport Island was created a few years ago from more than 500,000 cubic yards of sand dredged from the bottom of the bay. Today, it serves as a protected habitat for many endangered species, including the California least tern. (Perdue 2007)

Other restorative projects have also occurred just south of the bay in the Tijuana River Estuary, a protected habitat for the breeding of light-footed clapper rails and a number of other species. Sedimentary build-up, largely induced by humans, was noted at the estuary's outlet into the ocean, causing the water level in the area to rise unnaturally. Luckily, the habitat was mostly restored by dredging and greater damage was avoided. (Marcus 1989)

While good-intentioned dredging has successfully been used to restore habitats, it is not the ideal solution. Studies have shown that man-made and man-altered marshes do not function in the same way as natural marshes, and cannot achieve the same levels of productivity and biodiversity. While it is better for mankind to attempt to restore damaged coastal marsh habitats than to continue damaging them, it is far better still not to disturb nature in the first place, and to allow delicate habitats such as wetlands to thrive without our interference. Nearly all conservationists agree that the preferred strategy is to attack the root of ecological problems rather than focus too much energy in frantically repairing potentially irreversible damage. (Marcus 1989)

For better or for worse, the future of dredging seems clear. Because of human population growth and disproportionately large populations in coastal regions, dredging is still practiced across the globe and shows no sign of stopping. In fact, there are still many future plans in store for the dredging of San Diego Bay, such as converting many of the rocky, wall-like shorelines to more natural-looking sloping shores.

Most of the future changes that occur in the practices of dredging will likely be in the technological realm. There may be, for example, an increased use of suction dredges as pumps become more advanced. Other changes could result from legal action, with alterations in government policies regarding where and how often dredging may occur and what must be done with dredged spoils. While there exist many uncertainties regarding ecological conservation, one thing is certain: unless dredging is carried out in a more environmentally responsible manner, the damage that has been done to some of our most treasured coastal ecosystems could be irreversible. In the end, dredging has been and will continue to be an important and influential issue that directly impacts San Diego Bay and its inhabitants. (Perdue 2007)

The old Naval Training Center sits on dredge spoils along the Boat Channel. High Tech High is located in the center of the photo.

THE LAND HAS CHANGED

Where beautiful trees once stood
Now stand the countless monuments of a nation.
Rolling hills are leveled
And paved for the vehicles that traverse my surface.
Majestic mountains are robbed of their minerals
As the need for precious stones and building stones increases.
The growth of this nuisance that plagues my shoulders
Like a parasite, a disease that wishes to be a force of nature.
No one notices
The changes that they have brought to my land.

— Gabriel Del Valle

Mitch Perdue

As we waited at the Shelter Island boat dock for our interview with Mr. Perdue, we joked that as an important naval official, he would probably arrive by boat or submarine. Imagine our surprise when he actually came speeding towards the dock in a small marine research vessel and offered to take us on a tour of San Diego Bay, highlighting many of the important dredged areas. We raced across the bay at a speed of more than 40 knots, thinking about what a rare opportunity we were experiencing as Mr. Perdue began to share the incredible wealth of knowledge he has acquired in his years of work as the senior biologist for the U.S. Navy in the Pacific. At last we arrived at Homeport Island, created by the Navy from dredged spoils as a habitat for local endangered birds and fish. It was there we began our interview, literally standing on one of the most important dredge-related sites in San Diego.

Student Researcher (SR): What are some of the areas of San Diego Bay that have been most significantly dredged?

Mitch Perdue: We've developed a very workable management plan called the San Diego Bay Integrated Natural Resources Management Plan that we created in partnership with the Port of San Diego as well as our resource agencies. Through that process we identified pretty much the whole dredge history of the bay. When you look at it, the South Bay, south of Sweetwater, remains essentially intact. From the South Bay to the San Diego-Coronado Bay Bridge, there has been significant modification, not so much in shoreline but in depth, with the exception of the Naval Amphibious Base Coronado and some of the area around the Silver Strand. Then, of course, the bulk of the manipulation has been north of the bay bridge along the city front, North Island and, of course, the creation of Harbor Island and Shelter Island.

SR: How has geographic modification changed San Diego Bay in terms of economics, development and the environment?

Perdue: You will hear San Diego Bay being called a few different names: a harbor, a bay, a big bay and a working waterfront. It really is all of those things, with its three primary users being commercial, recreational and military interests. Dredging has been the key to every one of those.

Without the deepening of the harbor, we would have never been a naval homeport, and we could have never been what we are now—a mega-port. As a result of the military influence, we changed the bay bottom so that the commercial element can follow right behind. Although San Diego Bay will never be a major economic terminal like Long Beach or Los Angeles Harbors, we still have our share of commercial and maritime activity. As far as recreation goes, you really don't find any better bay than ours because we are a primary flat-water sailing area. You have to remember, too, that in the turn of the century through the thirties and forties, we were the primary tuna port in the Eastern Pacific.

SR: What are some of the dredging machines and methods that have been implemented in San Diego Bay?

Perdue: The dredger of choice, for most of the smaller projects in San Diego Bay, has really been the clamshell dredge. And that is just what the name implies: it's a grab bucket-type of concept that reaches down, grabs material, pulls it up and drops it on a barge.

On our larger projects, for channel dredging and aircraft carrier turning-basin projects, we use a hopper dredge or a suction dredge. If you can picture a vacuum cleaner sucking up the bottom, it takes up the material into a barge and dewaters it until the distillate (the water) spills over the side, leaving the sand remaining. This, although an effective form of dredging, has ramifications because of the amount of high turbidity it creates.

One of the deposits that makes up the bay bottom is called the Bay Point Formation, a geological formation of calcareous material, ancient shells and calcium deposits that are very hard. When we get down to this, we have to use what's called a cutter-head dredge, or even go back to a clamshell dredge, to break through that layer that can be two to six meters, depending on the location. The cutter-head dredge has big teeth and it can actually chew up the bottom. This material is then sucked up into the hopper and then transported off site.

SR: Could you describe the pollutants in dredged spoils, specifically, how they get there?

Perdue: When you think that, over time, what falls overboard from a ship or what falls overboard from a pier is pretty much unregulated through all these different periods—heavy military, commercial fisheries, the tuna industry, waterfront activities and kelp-forest harvesting—all of these activities have gone on, and the byproducts are found in the sediment layers, just like a cake. As these materials are dredged, the byproducts can be re-exposed to the water column.

SR: How have dredged spoils been handled over time?

Perdue: In the early part of dredging history you would take material from the bay bottom or the ocean and use that to create new land for building. Although great from an engineering standpoint, ecologically it was not so good. As laws like the Clean Water Act and the Coastal Act were implemented, what we moved towards was a "no net loss" concept of U.S. waters and "no net loss" of habitat. This meant more testing was required. Beforehand, the dredging program used to take about a month to prepare for, but now takes about two years.

SR: What have researchers or environmental agencies done today in order to reduce the spread of pollutants in dredged spoils?

Perdue: When we go down and we look at a site, we will take sediment corings and look at the whole coring all the way down to the depth to be dredged. We will do what is called a "green book" analysis of that sediment—looking at all the chemistry, all the volatiles, all the different types of target chemicals. From that analysis, which is a very arduous process that we do with the Army Corps of Engineers and EPA, we

will determine what is the most suitable action for that material. Some of it needs to be incinerated because it is so bad, some of it is so good that it can be brought back here (to Homeport Island) and used in a beneficial way; then we have everything in between. In other cases, we will actually leave that material in place in a confined site, because it would have less of an effect on the marine environment, while some of it absolutely must be removed.

A wide variety of material, such as this office chair, can fall into the bay and end up as dredge spoil.

SR: What are some of the dredging plans for the future of San Diego Bay and what are some of the ongoing projects?

Perdue: We are always dreaming up new and innovative ways to enhance the bay. Right now we are working on softening the bay shoreline to get more sandy, intertidal beach, versus the rock and armor stone. Our dilemma, not just with the Navy but with everyone, is that no one is going to give up that $1,000-per-square-foot real estate so that we can push the shoreline back and make it a nice sandy slope. We have to work with the engineers to come up with a configuration so that we can use armor stone or some other way to protect the upper part of the slope, while allowing the bottom end of the slope to remain sandy and function as a habitat.

Protected bird habitat of Homeport Island, created by dredge spoils in the bay, was the location for the interview with Mitch Perdue.

Land Use

"We've had so much industrial activity along the bayfront you really find toxicity in the sediments everywhere."
— Bruce Reznik

The coexistence of humans and wildlife in the San Diego region is thought to have begun 8,000 to 12,000 years ago. During much of this time, the land was used by prehistoric man solely as a resource for food, water and shelter. Little of the land was altered except for the foot paths that were used by these early settlers. Archaeological digs have unearthed several ancient sites around the coast, one being Ballast Point that revealed 6,000 years of occupation by Native Americans. (High Tech High 2007)

Because food was abundant in the region, the population thrived. This, in turn, led to land modifications to support the increasing numbers. Some of the earliest of these in San Diego may have been by the Kumeyaay people who redirected water to begin the first irrigation in the region. ("Kumeyaay History" 2007)

The development of the land at the cost of native wildlife was catalyzed by Juan Rodriguez Cabrillo, whose arrival during the Age of Exploration became the crowning achievement of Spain's efforts to expand into the New World. Word of Cabrillo's arrival reached Spain and would eventually bring about the voyage of another Spaniard, Sebastián Vizcaíno in 1602, who would later help establish the Spanish claim to the region. While minimal in comparison to the damages done today, it was Vizcaíno and his men who began to further develop the otherwise untouched land. ("Timeline of San Diego History" 2007)

The Spanish conquest would continue with the arrival of Gaspar de Portolá and Father Junipero Serra in 1769. These two men took the small Spanish encampment and expanded it into one of the most important Pacific ports. Portolá went on to become Baja California's first governor and oversaw the Spanish troops who were constructing a base in the region. Father Serra was made responsible for the construction of 21 Franciscan missions along El Camino Real— including the Mission San Diego de Alcalá. (McKeever 1985)

Portolá and Serra's efforts resulted in terrain that was once home to natural valleys and vast areas of coastal sage scrub being transformed into grounds for missionary buildings. One negative effect of the resulting population growth was raw pollution pouring into the bay. ("Timeline of San Diego History" 2007)

By the dawn of the 1800s, San Diego had become one of the few trade posts in the Pacific—a small seaport left behind by the Spanish colonists. However, winning the War of Independence against the Spanish rule empowered Mexicans to focus on agriculture and cattle-raising that would change the face of San Diego forever. ("Timeline of San Diego History" 2007)

As short lived as the Mexican era was, it brought change to the area: even more of the previously untouched land was used for the cultivation of crops and small, crude houses formed little villages. Large, communal cattle ranches began to appear throughout the San Diego region. The pristine land was forced to yield to the lifestyles of the Mexican rancheros and "cowboys."

The Mexican government encouraged this expansion, resulting in San Diego undergoing an unregulated agricultural revolution and

the bay's first major pollution problem. Animal feces used for fertilizer eventually ran off into the bay, and the lack of a sewage system compounded the problem as human waste also entered the bay's waters. (Mills 1985)

Mexican occupation of San Diego came to an end following the 1846–1848 Mexican-American War. In 1849, only a year after the United States acquired San Diego, the California Gold Rush began and people came from across the country—and in some cases, from across the world—to try and strike it rich in the California mountains. While gold mines were rare in San Diego County, mercury mining took place outside the city. Unfortunately, much of the mercury leaked into the lakes and groundwater that eventually ran into San Diego Bay. The seepage of highly toxic mercury caused massive damage to the bay's eelgrass beds. ("Timeline of San Diego History" 2007) ("UC Toxic News" 2000)

The Industrial Revolution followed, and the city expanded further to include areas such as Old Town and New Town, where whaling and fishing became the leading industries. Stretches of coastline were paved over to make way for the growing number of fishermen. By the 1880s, San Diego Bay had become one of the largest fishing ports in the Pacific. (May 1986)

The arrival of the railroad led to further industrialization, which inevitably had a negative impact on the environment. The 1880s and 1890s saw substantial residential and industrial growth in San Diego. The increase in developed land along the bay and farther inland was largely due to the construction of factories that produced mechanical parts, textiles and other goods. These changes forced development along the bayfront, reducing the amount of untouched land to a mere 25%. The thin border between the bay and the city also meant that pollution was much more common. San Diego's lack of environmental policy resulted in myriad pollutants finding their way into the bay's once pristine waters.

By the early 1900s, land in the San Diego area was developed further, not only for real estate but also for military purposes, such as a Navy coaling station at Point Loma. This activity caused further disruption to the native biodiversity. The U.S. Navy had become increasingly more prominent and it wasn't uncommon for San Diegans to spy naval ships out in the bay. President Theodore Roosevelt's Great White Fleet visited San Diego in 1908, dropping anchor off Coronado as the harbor was too shallow at that time.

In 1916, William Kettner, congressman for San Diego, proposed that the U.S. Navy consider constructing bases and housing along San Diego Bay. Kettner's proposal was approved by Franklin D. Roosevelt, then assistant secretary of the Navy. The growing naval presence prompted the decision to establish San Diego as the home base for the Pacific Fleet. (Perdue 2007)

During World War I, those plans were put on hold, and in San Diego factories that were once used to make machine parts and textiles now made gunpowder and artillery shells. The bay's port facilities played an essential role in the transport of military goods to the frontline in Europe. Following the end of the war, Roosevelt and Ketner's construction plans for Naval bases and housing began and, in 1921, the Naval Training Center (NTC) was built. (Shragge 2002)

To accommodate the new NTC buildings, the City of San Diego decided to expand into the bay, initiating the massive dredging that created the narrow waterway between NTC and the airport that still exists today. To accommodate deep-hulled Naval vessels, dredging increased the bay's water level to 17 meters and practically destroyed any traces of eelgrass in the process. At the same time, the sediment and metal deposits that leaked into the bay during the post-construction period poisoned and killed native fish. City expansion also included the Naval submarine facilities and the 32nd Street and North Island bases. ("Naval Training Center" 2007) (Shragge 2002)

The industrial-military complexes along the bay were creating heated, metal-contaminated water that poisoned sea life as it entered the bay. The economic boom that resulted from homeland manufacturing influenced the arrival of a period of economic prosperity known as the "Roaring Twenties." ("The Jazz Age: The American 1920s" 2007)

Similar to the real estate boom in the 1880s, the big-city lifestyle was gaining popularity among American youth and San Diego continued to expand as never before. The city soon became a popular tourist destination with famous landmarks such as Balboa Park. More land around the bay was being developed for commercial and recreational use. The real estate increase meant more money for the city, but there were few regulations on the location of houses or offices. ("Timeline of San Diego History" 2007)

In 1941, the attack on Pearl Harbor resulted in enlistments in the U.S. Navy. The NTC became one of the most active military bases in the nation, training more than 400 sailors a year, and new Naval housing such as the Linda Vista Naval Housing Unit was created. Factories returned to the production of military materiel while the City of San Diego further altered the bay to allow better access for incoming and outgoing Naval ships. The modification of the land in and around the bay destroyed the few patches of eelgrass and other natural habitats. (McKeever 1985) (Shragge 2002)

The end of World War II brought with it not only the baby-boomer generation, but freeways, industrialization and surburban sprawls. For San Diego County in the 1950s there was a significant surge in real estate construction from Escondido in the north to Chula

227

Vista in the south. As more people needed a car to travel to and from the city, automobiles became a necessity for many households; petroleum runoff was now a common pollutant entering the bay.

With the large population boom, there was a sudden rise in food-processing plants that, like the factories of World Wars I and II, constantly discharged waste into the bay. The pollution and land used by these plants caused the destruction of natural wetlands and shorelines, particularly the intertidal and subtidal habitat in south San Diego Bay. By the mid-1950s, more than 50 million gallons of raw sewage were being dumped directly in the bay daily; in 1955, the California Department of Health issued a quarantine of all San Diego beaches and shorelines.

Over time, what was once natural habitat for abundant invertebrates, fish and seabirds would become filled with dredge spoils and paved with asphalt. The City of San Diego began to push for the creation of "human playgrounds" such as Shelter Island—one of the many projects implemented by the City of San Diego to transform the Naval town into a prime tourist destination.

In the early 1960s, Rachel Carson's *Silent Spring* seemed to finally raise concern for the repercussions of many of mankind's irresponsible activities, and still today is often credited as the book that began the modern environmentalist movement. (Carson 2002) In 1972, the United States banned the use of the pesticide, DDT. Across the country, environmentalists gained support, inspiring people to focus on nature itself. ("UC IPM Online" 2007)

With the success of nature-oriented attractions, such as SeaWorld, the San Diego Zoo, the San Diego Wild Animal Park, and even the bay itself, tourists flocked by the millions each year to see "America's Finest City." By the early 1990s, the city's residential needs continued to grow. Condominiums and bayside housing were popular and real estate developers did all they could to supply the demand. By the end of the twentieth century, only an estimated 15 to 20% of the original coastline of San Diego still existed. The city government, which had already organized and set aside land divisions to safeguard natural spaces, faced constant pressure by developers to convert more space for human use.

However, in the 1990s, the Environmental Protection Agency (EPA) initiated several assessment reports of San Diego Bay and identified numerous environmental concerns. Many of these violations addressed the negative effects of land development—urban runoff and the destruction of wetlands and marshes along the San Diego coastline. The Port of San Diego and the five cities that border San Diego Bay (San Diego, Chula Vista, National City, Imperial Beach and Coronado) have continuously attempted to resolve these issues. These governmental agencies are all regulated under San Diego County

Municipal Separate Stormwater Sewer Systems (MS4) permit. (Port of San Diego 2007) (Reznik 2007). This Permit, instituted by the San Diego Regional Water Quality Control Board, describes the actions that governmental agencies must take to prevent urban runoff. Additionally, the Standard Urban Storm Water Mitigation Plan (SUSMP), a part of the overall San Diego MS4 Permit, has served to reduce pollution from new developments and extensive redevelopments by capturing and treating runoff onsite. ("Port SUSMP" 2007) (Reznik 2007)

In response to the MS4 Permit, every governmental entity in San Diego County developed a Jurisdictional Urban Runoff Management Plan (JURMP) that addresses how its respective agency plans to prevent stormwater pollution. Each JURMP is divided into various sections to address all relevant aspects of pollution prevention, such as public outreach, commercial inspections and the monitoring of development and construction. (Port of San Diego 2007[1]) (Reznik 2007)

At the same time, Total Maximum Daily Loads or TMDLs have served as restoration plans for already contaminated waters. TMDLs are calculations of the maximum amount of a pollutant that a body of water can receive while still meeting water quality standards; in some cases they serve to help control the chemical and garbage runoff already present or directly entering the bay. However, though nearly 100 waters were designated as contaminated, including many areas of the bay, only a handful of TMDLs have been adopted. (Reznik 2007)

A more straightforward solution than the calculation of TMDLs is simply to treat contaminated water. One company, Clear Creek Systems, treats stormdrain water, making it pure enough to be transported back into the ocean or another body of water. Unfortunately, this method is expensive, and rarely used. Most cities do not have adequate funding for a system like this, and therefore utilize the less expensive method of identifying where the pollution originates on land and trying to stop that source. Should cities begin to advocate the direct treatment of polluted water, as the treatment technology advances, it would incrementally become more and more affordable. (Gustaitis 2003)

According to a report in 2000, more than 85% of the coastline around San Diego Bay has been divided and devoted to construction—whether by the Navy, city, or private industry. As a way of addressing this problem, the Port of San Diego and the Navy collaborated to create the Integrated Natural Resources Management Plan (INRMP). According to the overview of the plan, "This INRMP provides the goals, objectives, and policy recommendations to guide planning, management, conservation, restoration, and enhancement of the San Diego Bay ecosystem. It also provides support to the U.S. Navy and the Port missions. As such, it serves as a nonregulatory guide to better, more cost-effective decisions by those involved with the bay." As part of the plan's implementation, development cannot occur along the bay without a permit from the Port of San Diego. This was done so that the bay's remaining wetlands and marshes would remain as natural ecosystems and provide wildlife with sanctuaries from development. (Port of San Diego 2007[2])

Another proposal is one of "Smart Growth" put forth by the San Diego Association of Governments. This plans for the future of the region and considers San Diego's continued human population growth while focusing on reducing urban sprawl, locating housing near employment centers, and providing people with alternative travel choices to reduce pollution from automobiles. Our local wildlife cohabitants may still have a chance with some better planning than was done in the past. ("Land Use—Regional Growth" 2007)

This heavily industrialized shore of San Diego Bay sits in close proximity to a federal wildlife refuge.

The Port of San Diego also focused their efforts to restore populations, as well as the habitats, of endangered species threatened by land use and pollution. In 1999, the Port established the South San Diego Bay National Wildlife Refuge as an extension of the work the Chula Vista Nature Center (CVNC) had done since 1987. Since then, both the CVNC and the Wildlife Refuge have worked towards preserving the remaining wetlands in South Bay, providing animals, whether endangered, threatened or common, a safe habitat away from human intrusion, pesticides and land development. More than 2,300 acres of the bayfront and surrounding wetlands are under federal protection; both the Navy and the Port are involved in the protection of these animals. (Port of San Diego 2007[3])

Unfortunately, the cities along San Diego Bay still find themselves pressured to expand. Everything from more housing and recreational places for tourists to freeways and power plants has been on council agendas numerous times. In many instances, land has been used and nature sacrificed for tourism and business. (Port of San Diego 2007[3])

A specific example of the tension between development and the environment is the replacement of the South Bay Power Plant in Chula Vista. For years, the City of Chula Vista and the Port of San Diego officials have been debating whether or not to allow a new plant to be built there. Although the current power plant produces more power than the one proposed, it causes air pollution and discharges heated water, which has harmed the aquatic environment near the plant. While several officials claim that the proposed plant is a necessity, the Chula Vista mayor and council stated that they did not want a power plant on their bayfront—old or new. (Mannes 2007) (Lee 2007)

The City of San Diego has also been pressured to expand the Point Loma sewer plant to better control and filter runoff going into local waterways. However, both the U.S. Navy and the National Park Service have denied the request for land because the expansion would intrude on federally protected land. At the same time, another battle is brewing concerning the expansion of a key part of San Diego—Lindbergh Field. At the Global Logistics Symposium of 2007, Port Commissioner Stephen Cushman proposed that the expansion of road access to the tarmac for logistical purposes and cargo use would boost San Diego's economy—but at a cost to the bay. (Lee 2007)

While it is encouraging that many city governments will make the right decision in matters that directly affect the environment, there is still concern that some officials do not feel that environmental policy is applicable. The mindset of the Industrial Revolution—the need to construct, change, and manipulate to continue our financial progress—appears to remain in the forefront.

San Diego has had a long history of human impact on the environment—most of it harmful. The human desire prevails for more cargo—today known as the accumulation of wealth—both for the city and its people. Our obsession with cargo and money has led us to where we are today, but the right education can help the world with environmental laws and better choices. To do this, we should continue to encourage and inform the Port of San Diego, the city councils, and state and federal governments to make more environmentally sound policies and decisions in the future.

AN UNFORTUNATE REVOLUTION

Specks of sun glistening on the bay
Wind blowing through the grass
Birds soaring through the sky
Construction sounds in the background.

Buildings going up everywhere.
Houses filling the land.
Economies over ecosystems.
That's what it's become.

Buildings and towers
Replacing the green.

— Shayna Goodman

Bruce Reznik

When considering individuals with expertise on the subject of man's impact upon the shores of San Diego Bay and its surrounding coast, Bruce Reznick's was one the first names to come to mind. Mr. Reznick is the Executive Director of San Diego Coastkeeper and is a champion of the environment.

As we walked over to San Diego Coastkeeper's office at Liberty Station, we found ourselves both anxious and excited to complete our first interview. When we met Mr. Reznik, his relaxed and personable demeanor quickly calmed our restless nerves. We began to scout for possible interview locations among the perfectly landscaped grounds. With the radiant sun above us, we came across a small courtyard with a fountain and knew instantly that it would the perfect spot to hear Mr. Reznick's story and his approach to environmental justice.

Student Researcher (SR): Where have you seen the most development pressures in San Diego Bay?

Bruce Reznik: I would say that the challenges are different throughout the bay. In South Bay, you have a lot of the last remaining open, natural space; the challenges in South Bay are, of course, the South Bay Power Plant as well as developmental pressures to build on the last undeveloped area. If you move northward up the bay near the Coronado Bridge, you've got the military facilities and shipyards. This area is already fairly developed, so you don't have to worry about a lot of development pressure.

SR: Can you explain the effects of development on the bay?

Reznik: Development is really the primary contributor to the largest problem we have on the bayfront, which is urban runoff. Even when you're looking at things like shipyards and industrial activities, a lot of the problem comes from "paving over," leading to the toxins going directly into the bay. The single greatest contributing factor to urban runoff is how much hardscape you have in the area.

SR: How have you tried to reduce these effects? Have you been successful?

Reznik: When we brought up a lawsuit against Caltrans over how they build roads, one of things we required was that they needed to try all these different treatment devices, or "Best Management Practices." At Caltrans, engineers use sand filters on all the runoff from roads to get cleaner water. But, as an alternative, if you just use something like a really simple bioswale [natural grassy swale], the water that infiltrates it filters out pollutants from runoff just as well and at a fraction of the cost.

One of things you learn is the closer we follow Mother Nature the better. Unfortunately we've lost a lot of battles along most of the bayfront, and even in the watersheds in San Diego Bay. The more we use something called "Low Impact Development" (or LID), the more we're thinking about natural systems and how we can best mimic them, so the better off we are going to be.

SR: Where have you found that San Diego Bay is the most polluted and where does it need the help the most?
Reznik: The pollution in San Diego Bay is pretty chronic .... I think most of the studies have shown that, and it makes sense. San Diego Bay is an enclosed bay; there is not a lot of "flushing," so pollution that goes into it tends to stay. We've had so much industrial activity along the bayfront; you really find toxicity in the sediments everywhere in the bay.

SR: How can the average person help to stop the current pollution in San Diego Bay?
Reznik: People need to recognize whatever we put in the streets—whatever goes into the storm drains—is coming out the other end untreated into the bay or into our coastal waters. So, we need to make sure nothing goes down storm drains that shouldn't be going down—basically nothing should be going down other than storm water. We need to make sure we're watching our pets, and maintaining our cars properly. I think it's important to have a voice in what we do. I think it's marching down to city hall, and I think it's writing editorials and letters to the *Union-Tribune* and really making sure that people recognize this is a priority .... It's been so frustrating over the last decade watching this country come closer to a cliff in terms of environmental protection. Nobody seems to want to talk about it.

SR: What pollutants have you found to be the most harmful to animals in the bay?
Reznik: I think they are the [pollutants] that last the longest. People tend to be very cognizant, frankly from a self-interest standpoint. I think it's really a lot of the man-made toxins like PCBs [polychlorinated biphenyls] or PAHs [polynuclear aromatic hydrocarbons]. A lot of the metals, once they get into the sediment, are there for generations or millennia, and a lot of these are carcinogens. They impact the fish population, and

as they move up through bioaccumulation or biomagnification, they pose a serious threat to people who fish along the bay. This is a country where, at one time, people got their sustenance from individual catches. They went out and they fished for a living, especially in bad economic times, and would catch fish to keep their family fed. Today, you still have sustenance fishers; it may not be something we are aware of, but if you go to the piers along San Diego Bay, you're going to see people fishing as their main source of food. Yet, what they're ingesting is mercury, PCBs, and PAHs. Those contaminations may have come from 40 to 50 years ago with our industrial activities, and they are going to last hundreds, thousands of years more.

SR: What animals have you noticed seem to be the most devastated by the pollution of the bay?

Reznik: It's so pervasive throughout the bay, I don't know if I can answer! We've changed the dynamic of San Diego Bay so much, we have a hard time looking at it as even a truly natural ecosystem. We literally have dead zones along the bay where there's nothing left; nothing living. Whether it's Navy facilities or the ship yards, we've had divers go down as part of our lawsuit against San Diego Bay, and they find nothing [living]. If you go out, you can see it for yourself. If you go out on a boat and you look at the piers, you don't even see barnacles around there because there's so much toxicity. It has really impacted every part of the bay; you can only imagine what the bay was like hundreds of years ago when it was a calving lagoon for the gray whale and other species. We'll probably never get back to that, but one of our goals should be to restore a natural ecosystem as much as we can.

THE EARTH'S VOICE

Long ago the earth spoke to us
Its voice could be heard in the air, ground, and water
It would speak and we would listen.

It gave us shelter and we thanked it.
It gave us food and we thanked it.
It gave us dominion and we abused it.

Now we refuse to listen.
We blindly continue.
Someone must hear its voice once more
And herald a new age.

— Ross Aitken

Climate Change

"We need leaders in every big country to take the climate problem seriously and make it a top priority."
— Michael Oppenheimer

The current environmental crisis in the world is climate change—rising CO_2 levels produce a gradual warming of the earth's atmosphere, and contribute to ice-cap melting, extreme coastal storms and higher oceanic temperatures. While the issue of global warming has caused conflict in international environmental policy and there are those who question its veracity, the general consensus is that climate change is the result of human activity.

The consequences of global warming present serious threats to the planet's survival. As the overall temperature rises on land and in the oceans, adverse weather changes are likely to occur throughout the globe. Storms will occur at increasing rates and with far greater intensity and weather patterns are likely to become more unpredictable. Plant and animal species will begin to die off or will be forced to relocate, thereby degrading or destroying entire ecosystems.

Global warming is attributed to the emission of greenhouse gases. Just as greenhouses trap heat from the sun, certain gases in the Earth's atmosphere encompass the globe and trap heat. These include carbon dioxide, methane, nitrous oxide, water vapor and ozone. While these occur naturally and play necessary roles in keeping the Earth's atmosphere balanced, problems occur when that balance is disturbed. When greenhouse gas levels increase, infrared rays are trapped within the atmosphere and cause the Earth's temperature to increase.

Carbon dioxide is produced naturally by living organisms. During the last 400,000 years, CO_2 levels never went beyond 280 parts per million. However, since the Industrial Revolution, human-produced CO_2 has caused the atmospheric CO_2 level to double. This is attributed to a variety of human-related factors including factory emissions, cars and burning of fossil fuels. For example, since the introduction of the automobile, the overall global temperature has warmed 0.7 degrees Celsius. (Oppenheimer 2007) (Karling 2001) (Roleeff 1997)

Scientists have discovered that following the sun's patterns is an excellent way to predict rapid climate change. By observing solar activity, they have been able to determine temperature and weather patterns. According to Dr. Michael Oppenheimer, the orbit of the Earth and its exposure to solar activity affects temperature and climate to some degree, but what continues that trend is CO_2 fluctuations. Oppenheimer thinks that CO_2 does not start change as much as catalyzing and driving it—the more CO_2 contained in the atmosphere, the faster the temperature changes. Even small shifts caused by solar activity may have severe results as CO_2 exponentially increases its impact by trapping heat.

Some scientists also think that with an increase in the Earth's temperature, the ocean's ability to consume CO_2 decreases. Since the ocean is the world's largest "consumer" of CO_2, this is a big problem. Many researchers theorize that the ocean's CO_2 intake has almost stopped. (Oppenheimer 2007)

Because of the complexity of the variables involved in climate prediction, some skeptics argue that both current and past computer models are not accurate. However, in the scientific world, those whose dispute global warming now belong to a minority. To date the U.S. government has not adopted a leadership role in the international efforts to address the problem. (Durkin 2007) ("Hot Politics of Global Warming" 2007)

During the Clinton Administration, the Kyoto Protocol—global legislation that set mandatory target emission reductions—was signed by the United States. However, it was never ratified by the U.S. Senate and without this, our participation in international management of global warming was nonbinding. The first president to address climate change was George W. Bush, who attended the Earth Summit in Rio de Janeiro. The majority of summit attendees, including the United States, signed the U.N. Framework Convention on Climate Change. However, the Bush administration's involvement with global warming has, to a large degree, been rhetorical—focusing on the "unknown" aspects of climate change. ("Hot Politics of Global Warming" 2007)

Political and moral issues associated with global warming and climate change have led to legislative battles because greenhouse gases produced by one country, state or city may affect the entire global community. States have begun to operate independently to implement the Kyoto Protocol without senate ratification. Some state regulations include cutting fuel use by 20% by 2020, car emissions by 30% as of 2009, and placing a state cap on industrial emissions. California is an excellent example of this state/community-based action with the creation of special climate-change legislation that zeros in on land development. Developers must go through a rigorous review process of land and buildings to assure that the impact on global warming is minimized. (Oppenheimer 2007)

Scientists who study global warming have stated with 90% certainty that human activities are at its root. The sea level has risen about 20 centimeters during the last 100 years and the world's oceans continue to warm with ice loss from Greenland glaciers, polar regions and the Alps. Since the 1950s, the upper layers of the tropical waters have risen by 0.5 degrees Celsius.

Dr. Michael Oppenheimer predicts that if the Greenland ice sheets continue to melt, the sea level could rise between 17 centimeters and 61 centimeters globally by 2100. If this does occur, it would be a disaster for coastal cities; much of San Diego County would be submerged within the millennium. (Oppenheimer 2007)

One of the most useful documents on the topic is the National Assessment on Global Climate Change that suggests U.S. Western cities and rural ecosystems have been and will continue to be altered. The U.S. Climate Change Science Program (USCCSP) believes that increased CO_2 levels and a drier climate will cause more fires, a loss in forest productivity, and a possible loss of ecosystems in alpine regions. In the West, ecosystems have already been invaded by many non-native species while native species struggle to adapt. ("Climate Change Impacts on the United States" 2007)

Local weather patterns will also be affected. Although short-term weather effects are hard to determine, San Diego has undergone abnormal climate patterns in recent years. In 2001–2002, San Diego received seven centimeters of rain—one of the driest years in the city's history. In 2003, the county went 181 days without rainfall—a significant factor in the local Cedar and Paradise wildfires. That record was bested in 2004 with 182 days of zero precipitation, but was followed by 57 centimeters the remainder of the season, making it one of the wettest since 1940. In July 2005, record high temperatures were recorded in the county. Climatologist Ray Bradley has suggested that while these immediate weather patterns cannot be directly attributed to global warming, climate changes will continue to produce erratic weather patterns of ever-increasing magnitude. (Krier 2007)

Climate change not only affects weather but habitats and, most importantly, species and ecosystems. Over 100 years ago, San Diego Bay was home to more than 1,000 gray whales. Today, many scientists believe the primary threat to coastal whales is climate change. Grays tend to move into Arctic waters during summer months but migrate to warmer Southern California waters around for calving. However, as waters warm, foods found in Arctic regions have been pushed even farther north. According to John Hildebrand's research and tracking, whales have begun to migrate later in the season and are not moving as far south. (Bell 2006)

Marine invertebrates have already been affected by the documented rise in sea levels. Paleontologist Scott Rugh has correlated data on small invertebrates found in and around San Diego Bay. According to deposit data, many local shell and snail species have become extinct because of rising sea levels or temperatures, or have moved to cooler regions. Rugh believes climate change may promote invasive species in coastal habitats as native species disappear. (Rugh 2007)

After studying advanced prediction models, global warming scientists have concluded that two degrees Celsius is the tipping point, i.e., if the temperature rises by approximately two degrees, significant climate changes such as tropical storms and sea-level rise will occur. To stay under this tipping point, major international emitters such as the United States and China must become proactive in the fight against global warming. At the beginning of 2007, President Bush introduced the "Twenty in Ten Plan" with a goal to cut gas usage by 20% in the next ten years. In addition, Bush has announced plans for a conference with the top CO_2 emitters of the world for discussions on greenhouse gas emissions and future plans for reduction. Armed with the knowledge that coal and fossil fuel use emits CO_2 in large amounts, the international community must commit to a "green" phase to slow atmospheric deterioration. (Oppenheimer 2007) (Landler 2007)

Solutions that can help reverse the process of climate change include alternative energy and carbon-capture storage—a method of extracting CO_2 after it is emitted and then storing it underground or offshore. While resourceful, the storage lifespan is around 100 years before the CO_2 begins to seep back into the atmosphere. Alternative energy sources such as wind and solar power are not as efficient as other power sources but are environmental friendly. California is a leader in this with wind and solar farms across the state. Michael Oppenheimer believes the technology to moderate global warming is available and that the only roadblock is public and government motivation. The simplest solution is the implementation of stricter regulations for cutting CO_2 emissions at national and international levels, a measure that has already succeeded in many U.S. states and countries throughout the world as a result of the Kyoto Protocol. (Oppenheimer 2007)

Hopefully, through the use of readily available modern technologies and strong, mandatory legislation on a global level, we can begin to reverse the damage already done by the world's dependence on fossil fuels. Failure to act is not an alternative.

Fossil fuel use in the bay leaves behind much more than CO_2 emissions.

Michael Oppenheimer

We first met Dr. Michael Oppenheimer at the San Diego Natural History Museum during one of his many lectures. His knowledge concerning the dangers of climate change mesmerized us, and we became interested in conducting a personal interview with him, knowing that it would be an incredible opportunity to hear from one of the world's foremost climate experts.

Dr. Oppenheimer is the Albert G. Milbank Professor of Geosciences and International Affairs at the Woodrow Wilson School of Public and International Affairs, and Department of Geosciences at Princeton University. In the late 1980s, he and a small group of other scientists were able to organize workshops under the auspices of the United Nations that helped bring about the U.N. Framework Convention on Climate Change and the Kyoto Protocol. He appeared on an episode of *The Colbert Report* in February 2007 and, only a few months after this interview was conducted, he was a participant in the Intergovernmental Panel on Climate Change (IPCC), a group that won the 2007 Nobel Peace Prize, an award the group shared with former Vice President Al Gore. Since then, Dr. Oppenheimer has worked with other environmental committees in efforts to combat global warming. Following several weeks of attempts to contact the busy Dr. Oppenheimer, we finally were able to set up an interview.

Student Researcher (SR): How does the IPCC differ from other organizations, and what power do they have to take action that others do not?

Michael Oppenheimer: IPCC is the *only* ongoing international scientific assessment process *on the climate questio*n. IPCC has no power per se, but is very influential as the only advisor to international climate negotiators.

SR: What kind of national and international political changes have to be made to help battle climate change, and keep us under the "two degree Celsius tipping point" that you often mention?

Oppenheimer: We need leaders of every big country to take the climate problem seriously and make it a top priority. Each country needs to address its own internal greenhouse gas limits, and also agree to a long-term international treaty, with implementation of 10-, 20- and 50-year goals.

SR: Can you give us a brief comparison of technological carbon reduction, to overall emissions cutting with current technology? For example, California seems to be doing well in terms of regulation and CO_2 cuts.

Oppenheimer: Current technology can offset about 20% of emissions at a cost saving, another 30% for a moderate cost. Beyond that, new technologies will be needed to reach cuts as much as 80%. There is ample time to do the latter if we would only take the former steps promptly.

SR: What specific data and research in terms of science (rather than politics) are you doing in the world of climate change? And what are some interesting things you have found in your research?

Oppenheimer: I am focused on the question of the stability of the West Antarctic ice sheets, which contain enough ice to raise sea level by about four to six meters. The latest findings are quite troubling in this regard. The risk that the ice sheet will disintegrate is low in this century but increases thereafter. But the outcome could be set in place by our emissions over the next few decades.

SR: The Bush administration has lately been "working with other countries and world powers to cut emissions." Do you feel this is a way to distract the public from the fact that the administration does not want to set a goal? What is the best path?

Oppenheimer: I think this is a distraction and the administration should focus on reaching agreement with other countries within the so-called post-Kyoto negotiations.

One potential impact of climate change—a drowned California least tern embryo following an astronomically high tide; a scenario that could become common in the event of rising sea levels.

Tony Haymet

It was a warm and sunny afternoon in the winter of 2007 when we arrived at the Birch Aquarium at Scripps to interview Dr. Tony Haymet regarding climate change and the aquarium's newest exhibit, "Feeling the Heat: The Climate Challenge." The exhibit showcases cutting-edge research done by local scientists working at Scripps Institution of Oceanography over the last 50 years. Dr. Haymet is the director of Scripps as well as UC San Diego's vice chancellor for marine sciences, dean of the Graduate School of Marine Sciences, and a professor of oceanography at Scripps.

Once we met with Dr. Haymet, we passed the ever-popular tidepool plaza high above Scripps Pier, saw the majestic coast of La Jolla Shores and began setting up our equipment. With the occasional sea breeze and background sounds of children, we proceeded to interview Dr. Haymet amidst these minor distractions. His extensive knowledge of climate change and the research being done by scientists at Scripps, including himself, have helped us shape our story on climate change and the future of our environment.

Student Researcher (SR): What is Scripps' main focus of research recently? What would you say you are shifting your attention towards?
Tony Haymet: We are a broad earth science institution that studies the solid earth, atmosphere and the ocean. Clearly we are very interested in climate change. It has been 50 years of hard work by our scientists to get climate change on the national and international agenda, and we are going to be capitalizing on that by doing more research in the areas that are not well understood, such as ocean acidity and the impacts of melting ice sheets.

SR: Can you tell us a little bit about the new climate change/global warming exhibit at Scripps, as well as some of the work and the research involved with it?
Haymet: It's something we are very proud of because it's not only a great exhibit for everybody in the community, but there's a heck of a lot of Scripps science over the last 50 years that's gone into it. I think it's a little different than most climate exhibits because approximately ten scientists have worked on it using their own data in the display. It also points to things we are doing more research on: acidity, glaciers, continuing to monitor CO_2, etc. There is some indication that perhaps

CO_2 is increasing at a rate even faster than we have estimated. So, it lays out the history, but also shows where we are going.

SR: As far as climate change goes, what are researchers at Scripps currently studying that concerns California?

Haymet: Some of our investigators have done a great job in applying climate science to California. We are very interested in the state of the snow pack in the Sierras; the fact that the snow pack may melt rapidly, or earlier in the spring, and how that affects the water supply for California. We are also studying wildfires. As California gets drier and the soil moisture is reduced, it can increase the probability of more serious wildfires. Another thing we're interested in is coastal erosion. As sea level rises and we see increases in storm frequency, we should be on alert for coastal erosion on California's beaches.

SR: How have other fields of science contributed to your research on climate change?

Haymet: Biology is a big area for us and these days it's hard to study earth science without stumbling on this climate change issue. We're currently looking at the way in which the ocean will adapt to changes in temperature as well as acidity. In some parts of the world where the hydrological cycle is driven harder, there will be more fresh water coming from big rivers so the oceans will be less saline than they have been in the past. These types of changes might impact the biology of the ocean in the future. We're also very interested in geology and seismology. Currently, we're studying the seismology around the Salton Sea, Lake Tahoe and off the coast of Santa Barbara.

SR: Personally, where do you want climate change research to go? Where do you feel it needs to go?

Haymet: My main concern is what is going to happen in the next five to 15 years, when we're going to need a few good ideas to reduce our CO_2 emissions and reduce the effects of climate change. I think if we begin to flatten our emissions now and continue to do so for the next five years, hopefully soon after that we can return to the 1990s level of emissions. If this trend continues, then in 20 to 30 years we can get back down to the emissions we saw around the 1960s.

In October 2003 and 2007, wildfires ravaged Southern California. San Miguel Mountain in the South Bay, shown below, was one of the affected areas.

STORY OF THE
BAIJI RIVER DOLPHIN

A group of High Tech High students attended a presentation at the Hubbs-SeaWorld Research Institute (HSWRI) on the last population survey of the Baiji River dolphin, which inhabits the Yangtze River in China. The presentation, funded by Anheuser-Busch, a number of nonprofit organizations and private individuals, was given by Dr. Brent Stewart, a researcher at HSWRI and one of the scientists invited to participate in the dolphin survey.

The expedition of approximately 50 people, organized by the Institute of Hydrology Baiji Research Team and funded by the Chinese government, had embarked on a 3,500-kilometer mission using several boats equipped with high-quality optical and underwater instruments. The party had started out full of hope, but attitudes turned grim with the realization that the prospects were grave for the animal. ("Yangtze Freshwater Dolphin Expedition" 2006) (Stewart 2007)

As Dr. Stewart gave his presentation, the students noted it was odd that all the photographs being shown were in black and white. Later, they found out that the images were actually full color, but only appeared monochromatic because of the heavy pollution in the Yangtze River. A feeling of dread settled upon them as they realized how tragic it was that such a unique species had been let go so easily. (Stewart 2007)

There have been three known marine mammal extinctions caused by human activity: the Caribbean monk seal, through centuries of hunting and breeding-ground disturbance; the Steller's sea cow, wiped out by Bering Sea explorers in the late 1700s; and the Baiji River dolphin (*Lipotes vexillifer*), which became extinct recently, possibly from dredging activity, overfishing and pollution.

This unique dolphin subspecies was declared functionally extinct during the writing of this book. Its story has many parallels to the imperiled species of San Diego Bay, so word of its extinction caused an emotional ripple through the student's research. They realized the connection between the Baiji's story and the issues regarding animals on the edge of extinction in and around the bay. It became apparent that many of the destructive variables influencing the Baiji were becoming universal.

In the early 1980s, the Yangtze River was reported to have more than 400 Baiji River dolphins in its waters, and prior to large-scale human interference, pods of dolphins could be seen in the Dongting and Poyang Lakes that branch off from the Yangtze. (Massicot 2006) However, these waters have since been declared uninhabited by the river dolphin. Between 1997 and 1999, there were only 17 official sightings; the last recorded one being in the Tongling and Dongting Lake areas in 2004.

This dolphin preferred the greater depths of fresh water with strong currents and sandbars, although it was often spotted in shallower waters when searching for fish and shrimp. It had a stocky body with a flexible neck that tapered into a thin, elongated snout. The animal also had small, rounded pectoral fins and a smaller dorsal fin. It was equipped with an acute sense of hearing that transmitted sound waves from its jaw to the inner ear (cochlea). However, its eyesight was poor and its eyes much smaller than those of other dolphins—no doubt these evolved over time as it had little use for good vision in the murky waters of its habitat.

Sadly, it was clear to see that the Baiji, as well another marine mammal in the Yangtze River—a finless porpoise—were victims of China's booming industrial economy. The conditions of the Yangtze River had become increasingly more polluted both by water and noise. Since the dolphin used echolocation to find food and communicate, the heavy boat traffic greatly interfered with its ability to perform these essential tasks. Additionally, boats would agitate polluted sediment, making the waters nearly impossible to navigate visually, especially for an animal with already poor eyesight, and probably making swimming difficult.

Members of the Yangtze River expedition that searched for the Baiji and finless dolphin included Dr. Robert Pittman of NOAA.

The population survey's expedition group also witnessed great numbers of barges dredging the river bottom for cement production. Fifty percent of the world's cement is generated from this activity, which heavily impacts fish numbers—the diet of the dolphin and other river mammals. While it is tragic, it is understandable that such a delicate and elusive species could not survive under the unnatural pressures created by mankind. (U.S. Geological Survey 2006) ("Yangtze Freshwater Dolphin Expedition" 2006) (Stewart 2007)

A major stumbling block for any dolphin population recovery attempt was the animal's low birth rate, producing only one calf roughly every two years. There have been attempts to raise the Baiji in captivity. In 1980, researchers at Institute of Hydrobiology in Wuhan, China rescued a male dolphin, named QiQi. Despite several attempts, QiQi was never able to breed successfully and he died in 2002. ("Yangtze Freshwater Dolphin Expedition" 2006)

Ship traffic along the Yangtze was likely a leading cause of the demise of the Baiji River dolphin.

The dolphin QiQi (pictured here) was held in captivity by the Institute of Hydrobiology in Wuhan but died in 2002.

Towards the end of his presentation, Dr. Stewart asked the all-important question, "What's next?" Many species are equally, if not more critically, at risk. He pleaded that we take steps to save them, for all species are intrinsically valuable to planet Earth. He mentioned that if we had taken the same initiative with the Baiji River dolphin that we have with species such as the giant panda, maybe the animal could have been saved. China is prospering, but this economic upswing may have contributed to the downfall of this creature.

SOLUTIONS

"The wrongs done to trees, wrongs of every sort, are done in the darkness of ignorance and unbelief, for when the light comes the heart of the people is always right."

— *John Muir*

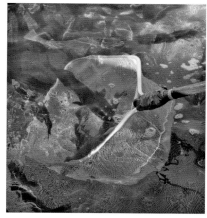

ENDANGERED SPECIES ACT
STEWARDS OF THE BAY
EELGRASS RESTORATION
SUSTAINABLE FISHERIES
MARINE PROTECTED AREAS
CARL HUBBS

Endangered Species Act

"It is important to have ethics—a shared moral dedication—to conserve the natural world."

— Rachel Muir

Some believe legal acts were needed much sooner to save the likes of the Carolina parakeet and the once prolific passenger pigeon. Those North American birds are gone forever, but we now have a law to protect species who need protecting. In 1973, a new and unique approach to the conservation of species was codified. This new concept was created with Congress rewriting the Endangered Species Preservation Act (ESPA) under the Nixon administration; this revised law was called the Endangered Species Act (ESA). With this new legislation, endangered and threatened species would be more effectively protected. The ESA allowed both plants and invertebrates alike to be listed as endangered, and provided protection by making it illegal to harm any listed species or its environment. (O'Toole 1996)

The ESA became and remains a powerful element of species preservation in the United States because, as former president Richard Nixon once said,

> "Nothing is more priceless and more worthy of preservation than the rich array of animal life with which our country has been blessed. It is a many-faceted treasure, of value to scholars, scientists, and nature lovers alike, and it forms a vital part of the heritage we all share as Americans." ("The Endangered Species Act" 2007)

The earlier law—ESPA—was created when the whooping crane neared extinction and congress recognized it was a significant issue. The U.S. Secretary of the Interior was authorized to create a list of endangered wildlife associated with the ESPA, and was given $15 million annually to create habitats for listed species. In 1969, the law was revised and broadened to include the protection of whales in the nation's waters. The secretary could also include foreign endangered species in the list and prohibit the importing of products made from them. (O'Toole 1996)

Even with these newly adopted powers, the ESPA was deemed inadequate. Consequently, with the ESA came many new tools to address and strengthen the laws regarding the world's threatened wildlife:

1. Simple Protection: Protection from hunting, which in the case of the American or Mississippi alligator (*Alligator mississippiensis*), was all that was needed to revitalize its population. In Mississippi, the alligator was hunted for its valuable hides and especially for its underbelly, which could be used in high-quality leathers. Unfortunately for the Mississippi alligator, unregulated and unrestricted hunting was rampant before the ESA came into effect. Today the alligator is a recovered species. (Sherry 1998)

2. Controlling Pollution:Banning dangerous pesticides, such as DDT, was a step that saved the majestic bald eagle (*Haliaeetus leucocephalus*) and several of the avian protagonists of this book. In 1947, when the bald eagle population dropped rapidly, it was discovered that DDT, a pesticide used to control mosquito populations, was the culprit. Unfortunately, fish ingested runoff containing DDT, and the eagle ate the fish. As was true for many other species, the pesticide weakened the eagle's eggs to the point where they would be crushed under the weight of the incubating parent. Since the banning of DDT on December 31, 1972, the

eagle population has risen from about 500 pairs in the 1960s to approximately 9,800 breeding pairs today. (Sherry 1998)

3. Critical Habitat Protection: Restricting most human activities in a specific habitat helped bring back the black-footed ferret (*Mustela nigripes*) from near extinction. In 1981, after the ferret was believed to be extinct, an isolated population of 129 animals was discovered in a prairie-dog colony outside Meeteetse, Wyoming. A number of the endangered ferrets were trapped by the Wyoming Game and Fish Department for a captive breeding program and after some years were released into a new location, the Conata Basin/Badlands Reintroduction Area of South Dakota. Human activity is restricted in this habitat so that the ferret population has a safer life. (Sherry 1998)

4. Conservation and Recovery Plans: These are formal statements of the tasks needed to re-establish a species. For example, the recovery plan for the black-footed ferret required breeding programs, the creation of a restricted habitat, an increase in free-ranging breeding adults, a reintroduction of the species into natural environments, and ongoing monitoring efforts, all before the year 2010. (Sherry 1998)

5. Captive Breeding: Individuals are captured in the wild and relocated to a protected facility where breeding programs are carried out. Because captive breeding requires controlled conditions for a species to prosper, it can be the most difficult means of species re-establishment but it can also be the most successful when dealing with an endangered species. An example of successful captive breeding would be the California condor (*Gymnogypus californianus*). In the 1980s there were only 21 birds remaining in the wild and in captivity combined. With the help of some captive breeding, in 1998 the bird's numbers had risen to 93 captive and 39 wild (released) condors, a dramatic increase in population. (Sherry 1998)

Causative agents used to evaluate the listing of species regulated by the Endangered Species Act (e.g., pollution, invasive species, loss of habitat, exploitation, climate change).

Despite the government's good intentions, there remain several main reasons for continued cause and concern surrounding Despite the government's good intentions, there remain several main reasons for continued cause and concern surrounding the ESA. First, between the years from 1973 to 1995, the ESA could not prevent the extinction of 108 species—an average of five species going extinct each year—meaning the nation's wildlife was clearly still at risk. One main cause for extinction was and remains that species were not added to the ESA listing quickly enough for them to benefit. Of the 108 species that became extinct during that period, 83 experienced long delays in gaining protection: 29 were extinct before being listed; 42 during a delay in the listing process, and 12 after a lengthy delay during which their numbers became too depleted to survive. (Suckling, Slack, and Nowicki 2004)

Second, the Department of the Interior faces conflicts with landowners over financial losses due to ESA regulations. For example, the endangered gray wolf (*Canis lupus*) is listed and their numbers have

The Carolina parakeet was once common in the Eastern United States and as far west as Nebraska. It was considered extinct in the 1920s.

now risen from just a few hundred to over 3,000. With this wolf population increase, livestock and other farm animals are being killed. However, a farmer or private landowner can be fined as much as $100,000 and face possible jail time for killing a protected animal. So, while the ESA benefits the wolf, it hurts many ranchers and farmers, and some rural Americans believe it is a poor trade-off. (Kostel 2008)

Third, it could be argued that the Department of the Interior does not have equitable distribution of its funding; more than 50% of its budget goes to just ten vertebrates. The uneven financial support of endangered species is an issue that needs to be addressesd. (Shrogen 2005)

Last, the public has lost interest in the ESA. Factors include: low media coverage; perceived lack of support by former administrations, and a sense of disconnect—the average U.S. citizen does not dwell on the fate of a threatened species over their own socio-economic concerns. Stories such as the imminent closure (at time of writing this book) of the Chula Vista Nature Center, a long-time supporter of endangered species, demonstrates the changing viewpoint of a local city government on the issue (see Stewards of the Bay chapter for more on this). However, even with this decline in public interest, the ESA still strives to protect endangered U.S. and foreign species.

As of February 28, 2007, the U.S. Fish and Wildlife Service (USFWS) listed approximately 1,350 species on its database, the Threatened and Endangered Species System (TESS). Unfortunately, TESS is purely for information. For a species to receive protection, funding and support necessary for rehabilitation, it must be listed under ESA. ("Working Together" 2007)

In San Diego Bay there are a number of different species that are not only in the TESS database, but are also protected under the ESA. One of these, discussed earlier in this book, is the Western snowy plover (*Charadrius alexandrinus nivosus*). This small shorebird is currently being monitored, as required by the USFWS. Fortunately for the species, it has a recovery plan that includes the restoration of critical habitat on the Pacific Coast. The plan requires the snowy plover to have 3,000 breeding adults for ten years, and an average annual productivity of at least one fledged chick per male. ("Species Profile: Western snowy plover" 2007) Another endangered species discussed earlier in this book is the light-footed clapper rail (*Rallus longirostris levipes*), a species of particular significance to San Diego that is currently receiving additional support because of its endangered status. This bird is a long-toed, long-legged, hen-sized marsh bird and was listed on the ESA on October 13,1970. The light-footed clapper rail has seven Habitat Conservation Plans (HCP) and one Safe Harbor Agreement (SHA), making a total of eight conservation plans dedicated to its rehabilitation. The clapper rail also receives additional support beyond that of the ESA from organizations including the Tijuana River National Estuarine Research Reserve. ("Endangered and Threatened Bird Species at the Tijuana Estuary" 2007) ("Species Profile: Light-footed clapper rail" 2007)

In addition to the federal ESA, there is also the California list of threatened and endangered species. This is similar to the ESA because both aim to protect fragile wildlife, create and monitor recovery plans, and work with organizations and landowners to benefit select species. Two plant species were added to the state list between 2001 and 2002: the Baja California birdbush (*Ornithostaphylos oppositifolia*) and the Orcutt's hazardia (*Hazardia orcuttii*). ("The Status of Rare ..." 2008)

There are many organizations that have formed to support endangered species. One is the Peregrine Fund, originally created to reverse the declining peregrine falcon (*Falco peregrinus*) population. This nonprofit organization helps birds-of-prey through captive breeding and other means. The Peregrine Fund works solely through private donations that are used in more than a dozen different national and international projects. One of these is restoring the California condor population in Arizona; the Peregrine Fund released and monitored 14 condors in 2005. The Peregrine Fund demonstrates how the general public can benefit endangered species through such programs. If the public were to become more aware of organizations like the Peregrine Fund, the crisis of declining populations might become a thing of the past. ("The Peregrine Fund ..." 2007)

The most likely change to the ESA in the future will be a provision that gives regulatory incentives to private land owners when they support a listed species. Under the SHA, Texas ranchers helped to restore the endangered Northern aplomado falcon (*Falco femoralis*) on their land. The process has also worked for the gopher tortoise, red-cockaded woodpecker, Schaus swallowtail butterfly, and a number of other endangered species. This is one way the ESA can

The snowy plover and least tern have protected nesting sites in San Diego.

be changed in the future to make it efficient enough to stay in effect. ("The Endangered Species: A Backgrounder" 2007)

Besides the ESA, there are many other ways in which endangered species can be helped; conservation research and increased environmental/ecological law enforcement are two of the most important. Conservation research will benefit endangered species not only directly, but through outreach efforts to the general public, especially its younger members who are the next generation of politicians, business owners and citizen activists. If people learn to help endangered species early in life, they can later help to influence public opinion to change the fate of many of the most at-risk plants and animals. An increase in law enforcement will help to protect endangered organisms by defining restrictions, limiting access to protected habitats, and otherwise reinforcing fines and sentences for those who harm the environment. With awareness and an enforced legal system, species protected by the ESA will have long-term positive outcomes and the potential to affect the nation's wildlife as a whole. (Ezcurra 2007)

Education is also important in helping our endangered species. If the United States and other nations around the world can teach the public about the impacts of their everyday lives, they will be able to understand how they are affecting the next generation. Only then can we have the reform that will lead to lasting, positive results. (Ezcurra 2007)

Rachel Muir

Over the course of our research, we were fortunate enough to come into contact with Rachel Muir, the Imperiled Species Coordinator of the U.S. Geological Survey. We could think of no better person than her to shed light on the subject of the Endangered Species Act (ESA) from firsthand experience. Muir is also the great-great-great-grand niece of the renowned naturalist, John Muir, the founder of the Sierra Club. She has followed the family tradition of conservation, endeavoring to preserve the vulnerable gems of America's biodiversity. Muir graciously received us at her office in Reston, Virginia and provided an incredible opportunity to hear from one of the most esteemed experts in the field of conservation.

Student Researcher (SR): How did your familial connection to John Muir impact you?

Rachel Muir: For starters, perhaps the long legs and strong back that have helped me explore the natural world are a gift from my ancestor! More likely is that his ideas—particularly the idea that all parts of the living and nonliving world are connected and bound together—have inspired me to understand and conserve the natural world.

SR: Can you explain the Imperiled Species Coordinator position?

Muir: Many kinds of taxa (species, subspecies and populations) of animals and plants are threatened with extinction. Many others have seen their numbers and their geographic distribution dramatically reduced because of a variety of threats. Most losses in species abundance and distribution are caused by human activities, such as changing habitats, (for example, forests to cities or wetlands to agriculture), introduction of invasive species, and exposure to contaminants, or climate change. As the Imperiled Species Coordinator, I assist our agency in directing and monitoring our research to find out what causes species declines and extinction and how they can be restored.

SR: What is an imperiled species?

Muir: It is one that has or is threatened with a biologically significant decline in population or geographic distribution. More properly, I am referring to imperiled biological resources; this includes species, and the finer taxonomic divisions of subspecies and populations, and the "big picture" level of habitats, biological communities and ecosystems. There are entire communities, such as the tall grass prairies of the Central

United States, or the mangrove forests of Florida and the Caribbean, that are nearly exterminated from their natural range.

SR: How do imperiled species relate to the Endangered Species Act?
Muir: "Imperiled" is a general term that, unlike "endangered" or "threatened," does not have a specific legal definition. It is a handy general term for species and habitats in significant decline. Not all species that are in serious decline actually get listed. We still know relatively little about the status of many species, the causes for their decline, and how they might be recovered. About 1,350 species are listed as threatened or endangered under ESA, but only a handful have been de-listed and fully recovered. Included among the species recovered are the bald eagle and American alligator.

SR: In your own words, what is the Endangered Species Act?
Muir: The ESA is a law that is one of the principal ways in which the U.S. government shapes its policies to protect endangered and threatened species of animals and plants and their habitats. The ESA has established a definition for threatened and endangered species, a procedure for listing them, and a process for de-listing them if they are fully recovered.

SR: Why is the ESA important?
Muir: It is one of the principal legal tools to protect biological diversity. Protecting all kinds of biodiversity is important—not just the handsome, feathered and furred creatures that attract the most attention. The ESA can be used to protect all species of animals or plants, (although it has never been evoked to protect a microorganism such as a bacteria or virus). Like the many parts of a complex machine such as a computer, each part has a role to play. What happens when you remove any part of a complex machine? Can it fail? How much more complex are the ecosystems on which we depend on for our lives and livelihoods! That is the importance of the ESA and other statutes that guide us in conserving our natural resources—the ESA helps us protect the parts of ecosystems called species. However, as you know, ecosystems are part of a complex mixture of living and nonliving things. And human beings as well! In my opinion, laws are helpful and necessary in guiding our activities and behavior. What is more important is that we have an ethic—a shared moral dedication—to conserve the natural world.

Stewards of the Bay

"It's about people standing up and drawing a line in the sand and saying we've got to leave this planet in better shape than we found it."

— Bruce Reznik

The term "steward of the bay" represents a broad concept. Stewards of the bay are not just defined as environmental organizations. It is such a broad term in that it encompasses everyone who makes a clear and consciousness effort to protect and conserve the environment, specifically San Diego Bay. At the beginning, individual recognition and effort fueled the environmental conservation movement. It was their conscious effort to restore the environment that led them to inspire others to action. The environmental movement, which became most notable during the 1970s, led groups of concerned citizens to establish local and national organizations to preserve the Earth's remaining natural resources. ("San Diego Bay Integrated National Resources Management Plan" 2000) ("State of the Bay" 2007)

Today, numerous organizations, ranging from nonprofit to federally run, are currently acting as environmental stewards for San Diego Bay. As stewards of the bay, these organizations have committed themselves to restore and sustain the natural resources and biodiversity of San Diego Bay for future generations. In the last forty years, stewards of the bay have created programs for the long-term protection and conservation of the bay's ecosystem by supporting education, research, land protection, and legislative initiatives. Thanks to stewards

of the bay, today's bay has seen much improvement since the days of its troubled past. (High Tech High 2005)

Human use of San Diego Bay dates back thousands of years to when the first coastal Native Americans used the bay and its tidelands as hunting and fishing grounds. Since then, San Diego Bay has dramatically changed. As settlers arrived in San Diego, the bay's natural resources were valued more for their potential human use: whales and birds for hunting, fish and shellfish for sport or commercial harvesting; wetlands and the bay itself, for dredging and landfill sites. During the twentieth century, the bay's health and biodiversity was greatly reduced due to the city's industrial boom. Sampling of the bay's waters during the 1950s and early 1960s revealed the deterioration. It was reported that by 1963 80% of the bay had dissolved oxygen concentrations below levels necessary for continued sustenance of fish and wildlife. (Browning, Speth, and Gayman 1973)

Over the past 40 years, the natural resources in San Diego Bay have experienced a significant resurgence. However, concerns have been raised about the future security of the bay's remaining habitats. As new pressures for more intensive use of the bay's shorelines and waters arise, development threatens the 176 bird species that are directly dependent on the bay. With an increasing number of invasive aquatic nuisance species also being introduced, the bay's ecological integrity is being called into question. Many agencies and organizations—on federal, state, and local levels—have been established to protect the bay's remaining resources. (Crooks 1997)

One of the first stewards in San Diego was the Port of San Diego. In 1962, the California legislature voted to create the Port of San Diego to manage the use of the tidelands surrounding the bay. Since then, the Port has been actively involved with the protection of

San Diego Bay's natural resources. Some of these resources include salt marsh and tidal flats, bird nesting and foraging sites, eelgrass beds, and nine federal and state listed endangered or threatened species. Over the years, the Port District has collaborated with other stewards to create management plans and programs that increase the impact of environmental stewardship in the bay. (Helly, et al. 2001)

Written in 2007, the State of the Bay (SOTB) is an effort of the Port to describe the current condition of San Diego Bay—its water and sediment quality, natural resources and habitat. The SOTB is intended to provide a "snapshot" of the bay, as it exists today. The SOTB evaluates the bay's condition based upon five criteria: physical features; habitats; water quality; sediment quality; and biology/ecology. To obtain the best overall assessment of the bay, as well as clarity as to where pollutants or impacts may be occurring, samplings were taken in four distinct ecoregions: north, north-coastal, south-coastal and south. The STOB is also a framework against which development or conservation decisions can be made for the bay. Because it was recently created, supplementation, as well as the implication of the SOTB, is only available when more data and research are conducted. ("State of the Bay" 2007)

Another prominent steward of the bay is the San Diego Audubon Society. The San Diego Audubon is a local chapter of the National Audubon Society, an organization founded in 1905 in honor of John James Audubon. The National Society was created to stop the slaughtering of birds for their feathers. Their early lawsuits helped to establish the Migratory Bird Act of 1913, which halted the plume trade and rescued many birds from the brink of extinction.

The San Diego chapter of the Audubon Society was established in 1917. Since then, more than 2,800 San Diegans have joined the society. Many of the members are avid bird enthusiasts, who participate in monthly birding trips. Over the last five years, the Audubon Society has grown from being an organization that only appreciates nature to an organization that is working actively to restore endangered birds in San Diego. Members have advocated for environmental conservation through education, as well as participated in numerous hands-on restoration projects. One unique society project supporting conservation research during the past century is its annual Christmas Bird Count. Since 1900, the organization and its volunteers have counted hundreds, if not thousands, of bird species throughout the United States. These counts give a baseline of the amount and diversity of species in a given region. ("SDAS Conservation Projects" 2007) Another prominent event of the San Diego chapter is the internationally recognized San Diego Bird Festival that attracts migrating bird enthusiasts from around the world during one week each March.

Pro Península is a bi-national nonprofit organization that works throughout Baja California. Conceived from a master's thesis by two students at the University of California San Diego in 2001, Pro Península has become one of the leading conservation groups in San Diego and Mexico. Throughout its short existence, Pro Península has employed researchers from both the United States and Mexico to promote and advocate for international environmental policy. Together, members of Pro Península have implemented and worked on various conservation projects throughout the region. As the organization has grown, its work has expanded from scientific research to environmental education and outreach, focusing primarily on sea turtles, water quality and ecosystem protection. ("About Pro Península" 2007)

Similar to Pro Península, Wildcoast works as a bi-national steward. Established in 2000, Wildcoast organizes conservation activities throughout Baja California and Southern California to protect ecologically important coastal wetlands, islands and marine areas. Wildcoast has focused its conservation efforts on four sites: Bahía de los Angeles, Laguna San Ignacio, Bahía Concepción and the U.S.-Mexico Border coastline. Since its creation, Wildcoast has successfully protected more than one million acres of coastal wildlands and halted the development plans of environmentally destructive tourist resorts throughout the Baja California Peninsula and the Sea of Cortez. Wildcoast has an environmental campaign called *Tortugas Marina* (Sea Turtles) that relies heavily on the media to convey its message. Since 2001, Wildcoast has employed a novel outreach strategy in conservation education by using celebrities such as bikini-clad Dorismar to advocate the nonconsumption of sea turtle meat and eggs. ("Wildcoast Annual Report" 2006)

In the early 1980s, the City of Chula Vista created the Chula Vista Nature Center (CVNC). According to Barbara Moore, a retired program manager at CVNC, "the Nature Center was created as mitigation—which means lessening the damage of the mid-bayfront development." (Moore 2006) Since its opening, the CVNC has become

an internationally recognized aquarium/zoo, which houses many plants and animals that are indigenous to San Diego Bay and its nearby wetlands. In 1988, the CVNC became incorporated into the Sweetwater Marsh National Wildlife Refuge. Since then, it has become one of the leaders, both nationally and internationally, on wetland-related programs and exhibits. It offers the public up-close experiences with various animals, ranging from the endangered light-footed clapper rails to sharks. (High Tech High 2005) ("About the Nature Center" 2002) As this book neared publication, potentially devastating cuts to the Chula Vista city budget may eliminate the Nature Center. Students and teachers involved in this study have participated in active protest and it is hoped that their efforts, along with those of others in the community, will save this amazing resource of biodiversity education and natural resource management.

San Diego Baykeeper, now known as San Diego Coastkeeper, was established in 1995 to combat pollution in San Diego Bay. The organization is the largest environmental advocacy group in San Diego. The organization also works in partnership with the internationally recognized Waterkeeper Alliance. As a local nonprofit organization, Coastkeeper organizes its conservation efforts around community-based programs. Programs include bi-monthly beach clean-ups with the Surfrider Foundation, large-scale community events such as Save the Bay 2005, managing the Harbor Safety Committee, monitoring watershed quality, and restoring kelp beds in and around San Diego Bay. Unlike other stewards of the bay, the organization is also an environmental watchdog group. Coastkeeper established a permanent law-policy clinic within its organization to fight environmental exploitation on a legal level. Over the years, the group, in partnership with other stewards of the bay, has filed over a dozen lawsuits against cities within San Diego County, shipyards, the U.S. military and Caltrans. ("Coastkeeper's Litigation Campaign" 2006) (Reznik 2007)

The Zoological Society of San Diego, founded in 1916 by Harry Wegeforth, is a nonprofit organization with more than 500,000 members—the largest membership of any such association in the world. It operates the San Diego Zoo, the Wild Animal Park, and the Center for Research of Endangered Species (CRES). The specific aim of CRES is to apply scientific innovation to the conservation and recovery of animals, plants and habitats, both locally and internationally. It includes five research divisions: Applied Animal Ecology, Behavioral Biology, Genetics, Wildlife Disease Laboratories, Reproductive Physiology, as well as a Conservation Education Lab. Over the last 30 years, the work of CRES has grown to include conservation programs in more than 20 countries worldwide. The center's success comes from the identification of the most pressing issues for species conservation and then applying the most innovative science possible to solving them. One innovation for conservation science is the collaboration between geneticists at CRES and the students involved in this study to do DNA barcoding experiments for species identification. ("CRES About Us" 2007)

Created in 1972, the California Coastal Commission (CCC) was established by voter initiative and later made permanent by the state legislature through the ratification of California Coastal Act of 1976. The CCC's mission is to protect, conserve, restore, and enhance California's coast and ocean for environmentally sustainable and prudent use by current and future generations. The Coastal Commission, in partnership with the City of San Diego, plans and regulates the use of land and water in coastal zones. Such development activities, like the construction of buildings, divisions of land, and other activities that change the intensity of the use of land or public access to coastal waters also falls under its jurisdiction. ("California Coastal Commission …" 2005)

The U.S. Fish and Wildlife Service (USFWS) is a branch of the U.S. Department of the Interior that is dedicated to managing and preserving wildlife. It began as the U.S. Commission on Fish and Fisheries in the Department of Commerce and the Division of Economic Ornithology and Mammalogy in the Department of Agriculture. It became the USFWS in 1939 when those bureaus merged and were transferred to the Department of the Interior. The goal of the USFWS is to conserve, protect and improve fish, wildlife, plants and their habitats for the benefit of the public. To do this, the USFWS has created and now manages more than 520 National Wildlife Refuges (NWR), including the San Diego NWR Complex. ("San Diego Bay National Wildlife Refuge Complex" 2008)

The California Department of Fish and Game (CDFG) is one of 15 Environment and Natural Resources Agencies in California. Similar to the USFWS, the department manages California's diverse fish, wildlife and plants and their habitats for their ecological values, and for their use and enjoyment by the public. The CDFG is responsible for the diversified use of fish and wildlife including recreational, commercial, scientific and educational uses. It has collaborated with local stewards, like the Port of San Diego, to create environmental impact statements that address development pressures on the bay, as well as possible solutions that benefit both wildlife and San Diegan residents. ("Department of Fish and Game" 2006)

Signed into law in 2003, Senate Bill 68 established the San Diego Bay Advisory Committee for Ecological Assessment (ACEA). The advisory committee was created to conduct an independent assessment of conditions and trends in the bay's health. Its findings were then published in the Senate Bill 68 report and used to supplement the creation of a comprehensive natural resources management plan. The report also established water-quality standards that were equitable to ensure the full protection of all beneficial uses of the bay.

The shoreline development of hotels, businesses and military facilities has significantly altered the natural condition of the bay's tidelands. As a result, initiatives were passed that protected the bay from further damage. Local, state, and federal stewards of the bay began to develop programs; many of these are collaborative efforts, which include a multi-step process to securely conserve the ecological integrity of the bay. Over the years the conservation movement has broadened from simple monthly beach clean-ups to public presentations by local experts on current conservation projects. ("San Diego Bay Integrated National Resources Management Plan" 2000) ("State of the Bay" 2007)

Complementing the Audubon Society's least tern program, the Port of San Diego's Endangered Species Management Program provides enhanced nesting and foraging opportunities for the endangered California least tern. This ongoing program creates a safe haven by restricting access to nesting sites and aiding with vegetation clearing and predator control. ("San Diego Bay National Wildlife Refuge Complex" 2008)

The San Diego NWR Complex, managed by the USFWS, was established to protect threatened and endangered species and their habitats, throughout San Diego and the bay. The San Diego NWR, consisting of the Sweetwater Marsh and South San Diego Bay unit, is located at the south end of San Diego Bay and encompasses approximately 2,620 acres of land and water. The 316-acre Sweetwater Marsh unit on the eastern edge of the bay supports the salt marsh and upland habitat, as well as the D Street Fill, an old dredge disposal site that provides nesting habitat for least terns and Western snowy plovers. The South San Diego Bay unit is approximately 2,300 acres and includes portions of the open bay, active salt ponds (i.e., Salt Works), and the western end of the Otay River drainage basin. The

Tijuana Slough NWR, a 1051-acre wetland including the Tijuana Estuary, is home to a variety of endangered birds. It is the only Southern California coastal lagoon that remains pristine, free of any roads or rail lines. ("San Diego Bay National Wildlife Refuge Complex" 2008)

In 1995, San Diego Coastkeeper addressed the pollution caused by local shipyards. Up against multi-million dollar companies, they threatened to sue shipyards, like General Dynamics NASSCO and Southwest Marine, over chronic stormwater permit violations, to address the seriousness of the matter. In 1997, NASSCO, the largest ship builder on the West Coast, now owned by General Dynamics, reached a settlement with Coastkeeper to conduct an environmental assessment of its facility, reduce contaminated runoff, and fund restoration of the California least tern and clapper rail nesting sites around the bay. In contrast to NASSCO, Southwest Marine fought Coastkeeper's threats. However, in 1999 the U.S. District Court ruled that the shipyard had to pay $799,000 in fines and ordered them to build a stormwater diversion system and increase the sampling and analysis of toxicity levels. Since then, Coastkeeper has partnered with other stewards to ensure that these shipyards continue to clean up the historic pollution they have caused. The shipyards along San Diego Bay are regulated under the California Industrial Storm Water Permit and their operations have no discharge to the bay. ("Coastkeeper's Litigation Campaign" 2006) (Reznik 2007)

The State Water Resources Control Board (SWRCB) was created by the California legislature in 1967. Along with nine Regional Water Quality Control Boards, including one in San Diego, the SWRCB's mission is to provide water quality protection and to develop and implement plans that best protect the beneficial uses of state waters. Together, the boards have restored more than 800 of California's pollution-impaired waterways to a condition level now safe for recreational use. The San Diego Regional Water Quality Control Board (SDRWQCB) has made great strides to restore the bay's health and has emerged as a model enforcer in California. In the past decade the Board has implemented various plans to restore natural habitats along the bay, improved watershed management throughout San Diego County, and pressured many bay users to clean up their pollution. Over the years, the agency has "earned the respect of many environmentalists while irking developers." (Rodgers 2007)

The SDRWQCB has become one of the most influential stewards of the bay. The board has worked on numerous campaigns to persuade boaters to avoid the use of copper-leaching hull paints, and with builders at over a thousand constructions sites countywide to install erosion barriers. Aside from campaigns, the agency, similar to Coastkeeper, sued the federal government to stop toxic sewage from being dumped through the South Bay outfall. The board's tactics to stop continual bay polluters has been so successful that in 2005 it was rated the best statewide agency for assessing penalties against violators. From 1998 to January 2006, the agency collected nearly $10.8 million in fines. Its most successful case was against the City of San Diego, in which the board collected an estimated $3.47 million in penalties. As a result, the number of sewage spills under the board's jurisdictions has dropped from six hundred spills in 1999 to 200 in 2005. (Rodgers 2007)

The Southern California Bight Regional Monitoring Program (SCBRMP) is a multidisciplinary effort coordinated by the Southern California Coastal Water Research Project (SCCWRP). The project's goal is to provide an integrated approach to environmental monitoring and evaluation of the region's coastal waters and sediments. The first regional survey was conducted in 1994 and involved 12 agencies that cooperatively sampled 261 sites along the continental shelf between the United States/Mexico border and Point Conception. The SCCWRP authorities and the other participating Southern California agencies intend to repeat these region-wide efforts every five years, with the next project scheduled for 2008. ("Southern California Bight ..." 2008)

Stewards have had to overcome countless hurdles to get where they are today. Often, they fight David and Goliath battles, whether about invasive species or continual bay polluters, and they face huge opponents when it comes to protecting San Diego Bay. Resources often come with limitations. Many of the stewards are nonprofit organizations that often rely on grants, federal funding and private donations to remain productive. It is difficult for them to have the resources or technical expertise on staff to work on the multitude of bay issues, but the challenge is these are needed to fight multi-billion dollar polluters, like NASSCO, and affect changes in the way the bay is treated. (Reznik 2007)

Another major obstacle stewards face when trying to preserve the bay is regulatory inefficiency. According to Bruce Reznik, an environmental attorney and executive director of Coastkeeper, "One of the most frustrating things is when regulatory agencies do their jobs very poorly and it ends up very hard to sue." (Reznik 2007) For stewards, it is easier to bring a lawsuit when regulatory agencies have done nothing to resolve the issues than when they are just doing their jobs poorly. The fact that these agencies study the problem and propose a solution shows that they are "addressing" the problem; however, oftentimes they fall back and propose other solutions that take more time than the environment's survival may have. Though these obstacles continue to affect the stewards today, they often have been able to overcome them to improve the health of San Diego Bay. (Reznik 2007)

There are more than three dozen public agencies that make decisions affecting the environment of San Diego Bay. These agencies have restored historically polluted areas and influenced policies to maintain the quality of the bay, which is healthier than it was 30 years ago. These improvements, however, do not mean the stewards' jobs are done. Stewards are now pushing lawmakers to see the value of retaining the ecological gains of the last decades and seek even more progress in pollution prevention, resource protection and balanced usage of the bay. ("San Diego Bay Integrated National Resources Management Plan" 2000)

As with many other coastal environments, San Diego Bay has been the subject of environmental monitoring activities by many groups over the past few decades. Monitoring programs are critical to assessing the health of the bay. While they have often provided useful information, some critical problems have occurred because programs have been too narrowly focused. These uncoordinated monitoring efforts have resulted in problems, including duplication of efforts, lack of standardized protocols and procedures for sampling and analysis, inconsistent data reporting, inaccessible data, and difficulty to the meaningful assessment and conclusion of the data. To fix this problem, stewards need to come together and coordinate a standard protocol. By standardizing protocol and using innovative technology, the stewards would be able to create better management plans for the bay. ("State of the Bay" 2007)

High Tech High student intern assisting the recovery of the light-footed clapper rail.

Bruce Reznik

Bruce Reznick was sought out not only for his expertise concerning the impact of land use in San Diego Bay, but also for his knowledge of local stewardship issues. His passion for environmental law developed early in his life, and Mr. Reznik's experience now makes him a formidable enemy to those who harm the environment. When he joined San Diego Coastkeeper in 1999, he started as the Executive Director, ready to direct the organization's efforts to protect and restore San Diego Bay. Under Mr. Reznik's leadership, the organization has grown to include eight full-time staff members and has done much to protect the environment through political action and education efforts.

Student Researcher (SR): Why do you believe environmental stewardship is important?

Bruce Reznik: Ultimately we have to live with what we do on this planet, and no one has to live with it more than the younger generation. We are the stewards of the planet, and you can't expect people to do your job for you. It started with individuals saying, "This is not acceptable. This is the water we drink, the air we breathe." I think that's really what Coastkeeper is all about.

SR: Could you talk a little bit about Coastkeeper and what Coastkeeper does?

Reznik: Coastkeeper is the largest environmental advocacy group in San Diego focused on coastal protection. We work in a number of different arenas. We do twice-monthly beach clean-ups as well as large-scale community clean-up events. We have an educational curriculum, water-monitoring program, and a kelp-restoration project. At the same time we also do advocacy work. We're a watchdog group. We go out and fight the city council and our state legislature. We push for laws. We hound people about pollution issues and ultimately, we sue people. In reality, that's probably not the majority of what we do but that's what we're most known for.

SR: How important is environmental law to stewardship?

Reznik: I believe environmental law is absolutely essential. There's a quote from the president of the Waterkeeper Alliance, Robert Kennedy Jr., where he talks about how corporate interest has taken over every form of government. We can no longer trust our elected officials to do the right thing because it's not a level playing field. The one place where we keep a level playing field is in front of the courts. Many of our largest and most influential victories are because we sued people. Whether we went to court or we settled, we've had huge impacts.

SR: What is the hardest part of protecting the bay?

Reznik: I think there are two things that make protecting the bay quite difficult. The first is the forces you're aligned against—our David and Goliath battles. When a group like Coastkeeper—even with a million dollar budget—fights against Southwest Marine and NASSCO, which is owned by General Dynamics, is difficult because these are multi-billion dollar corporations. We fight as best we can, and we try to do it as professionally as we can. I think we need more resources.

Regulatory inaction has been almost equally difficult. I look at San Diego Bay as a perfect example. We've known San Diego Bay as being toxic for 20 years. We know more or less the biggest polluters and they're multi-billion dollar corporations; they've got the money to clean up if they want. We do not have the political will through many of our elected officials and our regulatory agencies to do what needs to be done. The almost more challenging part is in environmental law because it's not that easy to bring a lawsuit. There are a lot of steps. You need to show that there is ongoing harm that is not easily correctable and that you're actually impacted; another thing you have to show is that the regulatory agencies aren't doing their jobs.

SR: What have been the most important factors in Coastkeeper's success?

Reznik: One of the reasons Coastkeeper is successful is the fact that we're a locally based movement. I believe it's a core mission for groups to be at the local level fighting to understand local communities. The fact that Coastkeeper is willing to use all the different tools in the toolbox is something important as well. It benefits the community and us. If you're not willing to draw a line in the sand and say, "This is pollution, this is harming," I don't think you can be effective. Lastly, I think the fact that we're a professional organization also makes us successful. We're fighting massive battles and if we don't have an infrastructure, professional staff, marine biologists and attorneys that we could work with, I question how effective we could really be.

SR: What is the most rewarding part of being an environmental steward?

Reznik: I'm blessed with waking up everyday getting to do something where I feel like I'm making a difference. There are so many different aspects of the job. I'm a policy wonk. I actually love going down to city council, going down to the courtroom, providing testimony, and winning a battle. At the same time, I love going out to a beach clean-up and working with community groups, going out and knowing that on a single day we're going to get debris off our beaches and [out of] our waterways. I'm really just lucky to do something that a lot of people often delegate to their spare time, and I get to do this everyday of the week.

SR: Can you tell us about how students, people our age, can get involved in stewardship itself?

Reznik: I think there are a number of ways, as far as what individual students can do to promote stewardship. The first thing is, of course, just to take care of daily activities. It's "don't put anything down the storm drain;" "conserve water and energy." Everybody could do all those types of daily activities, but I think it goes beyond that. The next thing, of course, is getting involved. Beach clean-ups are a perfect example. Most people say, "oh, it's a fun activity." But in reality, marine debris is a huge environmental problem. On an event like Coastal Clean-Up Day, in a single day in September, we get about 200,000 pounds of debris out of our waterways that would otherwise end up in our ocean, ultimately impacting marine life.

Bird biologists documenting a Western snowy plover nest in the Tijuana Estuary.

Eelgrass Restoration

"Our eelgrass has to be able to live in an environment where there is pollution of all different kinds."

— Dr. Kevin Hovel

In a few remaining locations of San Diego Bay, grass thickly covers the sea floor like a rainforest. A multitude of fish, including the giant kelp fish and sandbass, find this a safe haven where they can live and feast upon copious amounts of small invertebrates. This productive and diverse habitat is eelgrass. Unfortunately, eelgrass beds have been severely abridged or destroyed by dredging done to make way for the commercial development of San Diego Bay. A host of other factors have also led to the destruction of much of this once prolific habitat, and studies are now being conducted on restoration methods.

An important element in the ecosystem of the bay, eelgrass (*Zostera marina*) has both biological and geographical qualities that make it valuable to many dependent species. Eelgrass is not a grass, but rather an angiosperm. It is one of the few submerged flowering plants in the world and requires sunlight for photosynthesis. Typically, eelgrass blades are less than a meter in length and can be found anywhere from near the water's edge to as deep as seven meters below the surface, depending on the clarity of the water. Eelgrass plays many important roles in the environment, including providing habitat and food for numerous species, improving water quality and preventing erosion of the sea floor. ("San Diego Bay Integrated National Resources Management Plan" 2000) ("Eelgrass Update" 2007)

Eelgrass beds serve as both a refuge and nursery to epifauna, infauna, invertebrates, crustaceans and fish.

Epifauna and infauna are organisms that attach to the ocean floor or other parts of the marine environment to prevent being swept away by the current. If they are not attached to sediment, they find a home on eelgrass blades. Other species, mainly fish, use the eelgrass as a predatory refuge. Hiding among the leaves, these animals find safety for themselves and their young. In San Diego, young California halibut and lobsters are found in eelgrass beds, as well as the green sea turtle. ("San Diego Bay Integrated National Resources Management Plan" 2000) (Williams 2001)

Eelgrass is a primary producer and a foundation of the San Diego Bay food web. Many animals consume it as detritus (dead or decaying eelgrass bits floating in the water). Small invertebrates living within an eelgrass patch are consumed by crustaceans and juvenile fish. Fish living among the beds are eaten by fish-eating birds, such as the California least tern. Even waterfowl eat eelgrass as part of their dietary needs; the black brant, in particular, is one of the few birds that relies heavily on eelgrass as a dietary staple. ("San Diego Bay Integrated National Resources Management Plan" 2000)

Not only are the beds an important food source, they also improve water quality. The wavelike motion of the eelgrass blades slows the current, causing fine sediment and other particles to sink and eventually settle to the sea floor. The matrix formed by the roots of eelgrass beds beneath the sediment also prevents erosion of the sea floor. ("San Diego Bay Integrated National Resources Management Plan" 2000)

Right: Two-spotted octopus running to safety in a sparse eelgrass bed.
Below: Local shorebirds feast upon the organisms living on the eelgrass.

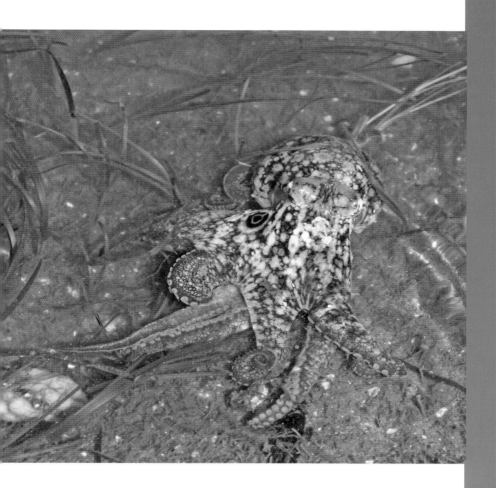

Though eelgrass in San Diego Bay was once abundant, it has since faced many risks and was even reported as eradicated. In 1941, there were 50,000 to 100,000 Pacific black brant geese living in the Spanish Bight, an inlet between Coronado and North Island. Because the geese feed on eelgrass, their numbers have been used as an indication of the abundance of eelgrass. After the Spanish Bight was filled, only 1,100 brant were reported living in the entire bay. ("San Diego Bay Integrated National Resources Management Plan" 2000)

Dredging has been one factor. As unwanted sediment is removed from a given area, acres of eelgrass can easily be uprooted and destroyed. Invasive animal and plant species pose further threats to the already sparse eelgrass population. One potential threat is *Caulerpa taxifolia*, which was commonly used as ornamentation in aquaria. This invasive seaweed grows very rapidly (approximately one centimeter per day), and can quickly monopolize an eelgrass area, effectively smothering the eelgrass to death. (Williams 2001)

The Asian mussel (*Musculista senhousia*) is an invasive species that directly impacts San Diego Bay's eelgrass population. Though it is not known when the mussel was introduced to the bay, it has grown so prolifically that its eradication is nearly impossible. Despite being small, the mussel forms dense colonies on the bay's floor; fragile, patchy eelgrass beds are the most vulnerable because they are unable to grow or expand. However, the Asian mussel dies in healthy eelgrass patches because its main food source, phytoplankton, is unavailable. (Williams 2000)

The other invasive species that has negatively affected the bay's eelgrass is the tropical stinging anemone (*Bunodeopsis*). The anemone population is so fast growing that during the summer it may cover entire blades of eelgrass, weighing them down and ultimately killing the bed. While this is a seasonal problem because the anemone population dies down during the wintertime, it does not disappear completely. (Williams 2000)

In San Diego, eelgrass also faces the less obvious threat of watercraft. Propellers and anchors can scar or remove small patches of eelgrass beds. Construction in the bay can also damage it. From 1999 to 2001, the retrofitting of the Coronado bridge required barges to be placed under the bridge where eelgrass grew abundantly. The eelgrass was severely damaged by both the barges' anchors and its obstruction of essential sunlight. Today, this sort of physical damage continues to pose the highest threat to eelgrass around the world. (Hovel 2007) ("San Diego—Coronado Bridge Seismic Retrofit Project" 2002)

Pollution from urban runoff has damaged the remaining eelgrass beds as well. Most things that go into storm drains provide nutrients that can be either beneficial or harmful. For example, algal blooms, which are a large source of oxygen, flourish from the nutrients. Yet these same blooms can drift and block the sunlight needed by eelgrass beds. (Hovel 2007)

Widespread dumping caused by sewage and industrial waste that polluted the bay for decades had adverse effects on the eelgrass population. However, restoration efforts were undertaken. The Point Loma Wastewater Treatment Plant and the Metropolitan Sewage System went into service in 1963, decreasing the amount of pollution discharged. With the restoration of water quality, eelgrass began growing again. Between 1994 and 2004 the acreage of eelgrass in the bay increased by nearly 43%, marking one of the most successful environmental comebacks in the bay's history. Today, eelgrass covers more than 1,600 acres. ("City of San Diego Water History" 2007) ("Integrated National Resources Management Plan" 2000) ("Management of Environmental Mitigation" 2006) ("State of the Bay" 2007) (Hovel 2007)

Eelgrass mitigation plans have helped to recover local beds, as well as other areas around the world. California's mitigation plan calls for not only restoring an eelgrass bed that has been damaged, but to replace it with more eelgrass than the original amount involved. Though in theory this plan seems effective, the restorers must decide whether to transplant already living eelgrass or to plant seeds, and neither method has proved to be easy. Often the success of these restorative efforts comes down to the location in which new eelgrass is planted. Transplanted eelgrass, especially in Southern California, tends to do well unless it is planted in areas that cannot support its growth, such as those that are too deep. ("Southern California Eelgrass Mitigation Policy" 1991) (Hovel 2007)

In San Diego Bay, there are distinct ecoregions in which different sets of aquatic species live: north, which is influenced by the ocean water that comes in with the tide; central, located around the San Diego-Coronado Bay Bridge; and south, which has low currents, warmer

Eelgrass acts as a substrate for a variety of invertebrates. Egg mass shown above was found at the Boat Channel and identified as bubble snail (*Bugula gouldiana*) from DNA barcoding in this study.

water, and less exchange with the ocean, perhaps causing fewer larvae to settle there. (Hovel 2007) These different ecoregions mean that transplanted eelgrass may not necessarily thrive in another area.

In September 2000, the Port of San Diego and the U.S. Navy Southwest Division released the Integrated Natural Resource Management Plan for San Diego Bay. The comprehensive action plan stressed the important role eelgrass plays in keeping a diverse and healthy aquatic environment in the bay. The Navy has now set up a mitigation site near the North Island Naval Station. The port's interest in the important ecological role of eelgrass was further expanded into a sponsored two-year evaluation, conducted by Drs. Kevin Hovel and Todd Anderson on eelgrass restoration. Their 2005 study used artificial eelgrass beds for monitoring different structural variables to identify those most important in making a habitat effective. Much of their study was conducted in San Diego Bay, where they looked at blade lengths as well as bed density. The researchers concluded that the most effective habitat for aquatic species depended on the amount of eelgrass present in a given area, rather than shoot density or length. (Natural Resources 2007) (Anderson and Hovel 2005) (Hovel 2007)

When looking at past experiences, it seems action only takes place when a need is brought to the public's attention. The future of eelgrass, therefore, depends on the city's awareness of the issue. The Port of San Diego, along with many other San Diego organizations, has the power and funds to take eelgrass restoration to its next phase. However, if its citizens are apathetic to the cause, no urgency will be relayed to the city and restoration may be forgotten. Thus, eelgrass restoration efforts prove to be another decision for the citizens of San Diego and around the world. Without protection for eelgrass, widespread development and ship traffic will continue to increase and flourish, bringing with it both industrial benefits and environmental drawbacks. Just as with all issues of conservation and restoration, its fate lies with the people. What is more important? ("Southern California Eelgrass Mitigation Policy" 1991)

Testing how eelgrass structure influences predator/prey behavior at San Diego State University's Coastal Waters Laboratory.

Kevin Hovel

We interviewed Dr. Kevin Hovel at San Diego State University's (SDSU) Coastal and Marine Institute near the bay. Dr. Hovel, the local "go to" expert on eelgrass, is an associate professor at SDSU. He has taught classes in marine ecology, statistics, conservation biology and other subjects since 2001, but the study of eelgrass as a habitat has been a major research project for him for many years. Talking about species that depend upon eelgrass for their survival, we took a tour of the deep basins filled with ribbons that simulated eelgrass. Finding a suitable place for an interview was an easy task amid the research going on in the laboratory. As we set up our equipment, Dr. Hovel opened our eyes to the vital and fragile world of eelgrass restoration.

Student Researcher (SR): Why is eelgrass important to San Diego Bay?
Kevin Hovel: It is the most important habitat that we have for [many] organisms that inhabit the bay. Eelgrass and other sea grasses form a structured habitat for a lot of species, particularly in their juvenile stages. Eelgrass is the most abundant shallow water habitat for a lot of invertebrates such as shrimp and crabs, and is also home to lots of the fish that people like, such as sand bass, giant kelp fish and kelp bass.

SR: How many native animals depend on eelgrass beds?
Hovel: That is a tough one. It is certainly in the hundreds or maybe the thousands. In fact, some of the work we do actually involves quantifying how many species use eelgrass. We take small samples of eelgrass and what we often find is dozens or more different species in one little area. So, it would be difficult to add up—with any accuracy—how many species use all the bay's eelgrass. We also have different ecoregions in San Diego Bay so species diversity and abundance differs from the northern to the central areas.

Eelgrass protects many species from predators. It is a great hiding place! One of the things I studied is how eelgrass functions as a refuge from predators. Also we have studied the effects of patch size; does the size of a seagrass patch make a difference in how the predators hunt in the beds, and whether or not they are successful in catching their prey? Eelgrass beds also serve as a source of food, though most species

285

actually do not eat the eelgrass itself, but rather what grows in the beds or lives in them.

SR: What would you say is the biggest threat to eelgrass in the bay?

Hovel: … I would have to say probably the amount of activity that goes on in the bay in terms of construction and shipping. For example, when the Coronado bridge was retrofitted by the state for seismic purposes, the state needed to place barges in the water under the bridge where eelgrass is abundant. The eelgrass was damaged by being in the shade of the barges and from their anchor chains that swept to and fro with wave action. That type of physical disturbance is probably the biggest threat to eelgrass and it comes not only with activities such as construction, but also jet skiing and boats that go into shallow water. Propellers or jets will scar the eelgrass; in fact, scarring the eelgrass—removal of small areas due to the disturbance that comes from the propeller—is an extremely common occurrence. It is probably worldwide one of the biggest threats to eelgrass.

Another big threat that we have here is actually eutrophication from sewage spills in the bay. At first glance, it might seem that nutrients might be good for any plant that grows in the bay. But what nutrients do is actually increase the abundance of algae, particularly macroalgae like sea lettuce; and when they grow quickly, they smother the eelgrass and wind up drifting to the surface and shading it. Eelgrass is one of those species that needs a lot of light and once shaded, it cannot photosynthesize anymore and dies.

Pacific black brant migrate thousands of miles down the coast from Alaska and stop over in San Diego Bay to eat eelgrass during the winter months.

SR: Are current eelgrass mitigation plans effective?

Hovel: Generally, my answer would be "yes" because it is actually a progressive plan. There are areas around the world where sea grasses are valued and are protected, but here in California we actually have a nice comprehensive plan that protects eelgrass more than in other places. Our plan basically calls for people to not only pay for damage that they do to eelgrass, but actually pay to restore a damaged bed and replace with more eelgrass than they took out.

Biologists like me sometimes have to determine how much damage has been done to an eelgrass bed by construction or pollution.

What that usually means is going in before construction actually takes place and assessing the amount of eelgrass and then going in afterwards to see how much has been lost. Then a report is submitted to an agency, like the Port of San Diego or Caltrans, and the company or whoever it is that did the damage. But with any mitigation there are always some issues. For example, if damage is done in one area, the eelgrass may be restored or put back into another area. However, our bay has different ecoregions as you go from north to south, which means if eelgrass is damaged in one area it might not grow well in another.

SR: What is unique about the eelgrass that grows in San Diego Bay?
Hovel: With our eelgrass it is the surrounding environment. San Diego Bay is like no other place where I have worked—it's an industrial, metropolitan area. A lot of other places you work in an eelgrass bed in the middle of nowhere. Here there are helicopters flying overhead and aircraft carriers coming in and out, and that means that there is a lot of disturbance taking place. So our eelgrass has to be able to live in an environment where there is pollution of all different kinds, lots of disturbance from boats and construction, and nutrient input and introduced species that reduce eelgrass health.

SR: How can the community get involved in helping eelgrass?
Hovel: I would say the community could get involved by being conscious of their actions and how these may influence the health of the bay. This could probably be the answer to a lot of coastal area problems. One of the things that damages eelgrass would be pollution—especially a lot of the nutrients that people put into the water. People do things that lead to lots of runoff going into the bay, and runoff can damage eelgrass and introduce species that kill it.

If people are boaters, they should also be conscious about potential damage to eelgrass. For example, when throwing down an anchor in shallow water, it might land in an eelgrass bed, and when that anchor is pulled up, a lot of eelgrass might be ripped out or scarred. Also, the eelgrass might be scarred by the boat's propeller. The fact that this is a very delicate, important shallow-water habitat, means that we have to be very aware about our actions in this bay that we enjoy so much.

Sustainable Fisheries

"If you had sustainable fishing, there would not be a problem .... You need to bring a lot of other things to the table to reach that point."

— Heidi Dewar

Fishing has been a practice of mankind for millennia. We harvested fish and other marine life from the bountiful waters as a means of survival, and by accessing this supply of nutrition, we became part of nature's food chain. As a species, we took no more to live and grow than was necessary. Nature then replenished this supply so that we are able to do so again, allowing each element of the cycle to continue. This careful interaction between predator and prey has been in place until relatively recently when the relationship changed. By disrupting that equilibrium, man has altered the way nature functions on a dramatic scale.

While primitive civilizations foraged, gathered and fished for sustenance and survival, we now harvest and exploit our planet for profit. With the development of agriculture and animal husbandry, the way humans lived underwent a radical change. Instead of living in symbiosis with nature, we now manipulate it to our advantage to maintain a world of surplus, separation and dominance. (Diamond 1997)

In most cases, profit is the objective above all else in the modern economy. Material worth of the corporation and individual rules; those without capital success are considered to be without power or social standing. This constant struggle for material gain may erupt into warfare and the subjugation of others as one group attempts to take advantage over another. We have created an economy based upon the concept of surplus and upon the human desire to accumulate material wealth beyond anything truly necessary for survival. This desire for "cargo" has massively altered our role on this planet, and the very balance of life. (Diamond 1997)

For years, corporate and profit-based fishing companies have regularly increased their harvests for greater earnings, contributing to population declines in many harvested aquatic species.

As fewer animals are left to reproduce for future harvest, we find their numbers dwindling over time. Fish account for approximately 16 to 29% of the animal protein consumed by humans worldwide. Since 1997, fish as a food source has increased by more than 32 million tons. Such a dramatic increase is beyond standards of sustainability, especially when considering the growth and expansion of human population. (Speer et al. 1997)

In San Diego, the situation is no different. During the late

For a period of time, San Diego was known as the tuna-fishing capital of the world.

1920s and early 1930s, the city became known as the "tuna capital of the world" when it entered the yellowfin tuna fishery. Being one of the two major points in tuna migratory patterns, San Diego was a prime location, and many fishing cultures chose to settle here. Massive catches were hauled into port every day. Tuna canneries soon appeared along the bayfront reaping profits from the harvest; several, like Bumblebee™ and Starkist™, were thriving businesses. The tuna fishing industry boomed in San Diego until 1976, when tuna numbers began a steady decline. Because of a need for "dolphin-safe" tuna, a growing population, technological advancements and demand, local tuna fishing collapsed along with a decline in fish stocks. (Brown 2001)

Awareness of overfishing grew, and by 2007 the tuna population had recovered in part, although numbers were not nearly as plentiful

San Diego is a world leader in the sportfishing industry.

or healthy as in the past. The decline in local abundance has been attributed to both overfishing and a shift in tuna migration patterns. (High Tech High 2007)

The white seabass was once a plentiful species in San Diego waters, but it was nearly driven to extinction during the last century. As a species with great commercial value, its overfishing, pollution, developments in commercial fishing technology and habitat loss have led to a drastic decline similar to that of the tuna. In recent years, efforts to replenish the white seabass have sprung into action in San Diego. Between education, in-water assisted growth (a form of restoration aquaculture), and efforts by local fishermen, the fish is returning to San Diego waters. (Rodgers 2006)

One of the most talked-about solutions to overfishing that might lead to our ocean's resurrection is sustainable fishing, a concept of careful balance. Early coastal and island-dwelling civilizations, whose cultures were heavily dependant upon fish stocks, took only enough to feed themselves and no more, leaving enough fish in the water to repopulate. Each year the cycle would continue with man playing an essential role in a balanced ecosystem. (Blackburn and Anderson 1993)

If commercial and sportsfishing today were to apply similar methods, our oceans might be helped. By leaving a sufficient number of fish in the waters to effectively repopulate, we might not exhaust this resource. This is a difficult task because of our population growth and continued desire for cargo and profit, and we may have to resort to other ways to help our oceans and still enjoy a seafood dinner.

One such technique is aquaculture whereby a species of fish is raised, maintained and harvested in a controlled environment. This method uses man-made enclosures, most commonly in open water, in which various species of fish are bred and farmed for commercial sale. Alternatively, farm-raised fish may be used to repopulate a species by being released into native waters. Either way, aquaculture plays a role in helping population numbers to revive and is becoming increasingly effective. With proper integration, we may be able to repopulate dwindling species. ("NOAA Aquaculture Program" 2007)

However, contrary to its appeal, aquaculture does have drawbacks. If a species is part of an aquaculture project in an area where it does not occur naturally, there can be drastic effects. Nonnative species in high concentrations may produce certain chemicals either through gland excretion or waste matter that could be potentially harmful to the surrounding environment. This issue must be taken into consideration when the ultimate goal is population recovery. ("NOAA Aquaculture Program" 2007)

This fishing seiner is followed home by a flock of birds.
Another boatload of fish for the market.

Another method currently being tested is called ecolabeling. This is a developing system in which commercially sold fish are labeled so concerned buyers are provided with information on their status. The shopper reviews the color-coded card indicating the fish numbers available world-wide. If the numbers are low, the buyer will hopefully reconsider that species and decide on another, more plentiful type of fish for dinner. This psychological approach to solving the issue relies upon global awareness to address the problem in our oceans. ("Introduction to Eco-Labeling" 2004)

To achieve sustainable fishing, sacrifices will have to be made. If we harvest an optimal amount while still reserving some fish for replenishment, the populations of fish might increase slowly and our goal eventually would be achieved. However, for repopulation at a faster rate, we would probably have to consume less than we would normally. In this scenario, the numbers of fish would increase faster, meaning more fish sooner.

The true question in this is where our priorities lie. It is a question of where the human race is choosing to go in the future. The fine balance between supply and demand cannot be maintained as in the past. The concept that supply is inexhaustible is no longer true. Our closed thinking has all but eradicated the wilderness. We could choose to continue on as we have for nearly a century, or we could choose to do something about the problem we have created, effectively taking our future and fate into our own grasp. To do so, we will have to step out of our comfort zone for a moment and address the world as it truly is—the fishing economy is falling at an exponential rate. There is absolutely no debate that we have a definite problem on our hands. However, it is entirely up to the world to decide whether something is to be done about it. (Eilperin 2006)

Until 2007, kelp had been harvested for 76 years just off the coast of San Diego by a company named Kelco (later ISP). Kelp extracts were used for a variety of products, from paint to ice cream. Kelco was praised for its sustainable practices, only harvesting kelp that grew within a meter of the water's surface. The sale of algin products extracted from harvested kelp was a highly profitable business, contributing about twenty million dollars to San Diego's economy annually. Though ISP Alginates no longer operates in San Diego, having in 2007 shifted business exclusively to its Scottish facilities, the company can still serve as an example of sustainability to which some fisheries and other marine harvesting businesses should strive. (High Tech High 2007)

This kelp cutter is docked in the bay, a reminder of the kelp industry that once boomed in San Diego.

However, with great awareness and a good measure of conservation, the dire situation in the ocean may potentially be saved. With multiple options at our disposal, we have a fighting chance to defeat our own ignorance and pride using the same inherent intelligence that got us to this point. This is the next step in our mental evolution. We have to take the step collectively and with flawless timing. If we are too late to save ourselves from our own mistakes, we may very well find the remainder of our stay here on this planet a considerably less pleasant one. We will simply have to wait and see how the world progresses in the future, and hope that we are smart enough to defeat the inherent flaws of our own intelligence.

Heidi Dewar

From high above the Pacific Ocean, standing on the balcony of the Southwest Fisheries Science Center in the winter of 2007, we could look out on the domain of Dr. Heidi Dewar. We had sought out Dr. Dewar after reading some of her studies on tuna physiology, behavior and ecology. As we began our interview we quickly realized we had found a significant steward of the fisheries. Dr. Dewar's calm nature, compassion for marine life and breadth of knowledge of fisheries and fish biology demonstrated why she is a lead fisheries research biologist for the National Oceanic and Atmospheric Administration. We also grew envious of her workplace environment, situated on the high bluffs of the La Jolla coastline.

Student Researcher (SR): How are sustainable fisheries related to endangered species?

Heidi Dewar: Most people think of sustainable fish as pertaining to the resource itself. When you are sustainable fishing for tuna, the tuna can replace themselves, but the problem is that the fisheries take more than just tuna. So, when you factor in the take of turtles, for example, it gets a lot more complicated. Fishing levels are sometimes set below sustainable levels to protect certain species.

SR: Is enough being done to support sustainable fishing?

Dewar: We definitely could use more money for research, and I think more needs to be done as far as public education and awareness goes; so many of the changes require the will of the people and some political force behind them. A better understanding of economics and how to best allocate resources would also help.

SR: What do you think is the best way for people to become aware of destructive fishing practices?

Dewar: That is a really good question and not an easy one to answer because in this day and age when people are just inundated with information from so many sources, it is hard to get any mind space at all. I think scientists need to be a little more creative about how they reach people—maybe use YouTube and blogs to reach the next generation. Certainly more PR for ocean issues would be great.

SR: To what do you attribute the decline of fisheries?

Dewar: There is always the tragedy of the commons where people do not act for long-term sustainability—"if they don't get it, someone else will, so there is no point in saving it for tomorrow." Also there are too many fishing vessels—a lot of over capacity—in some fisheries. Another factor is that often political elements come into determining what levels are set for certain fisheries rather than just the science or the biology. That is where increasing political will would help.

SR: Do you think sustainable fishing could solve the problem?

Dewar: That is somewhat circular because if you had sustainable fishing there would not be a problem. But you need to bring a lot of other things to the table to reach the point where you actually have sustainable fishing.

SR: Do you know of any examples of sustainable fishing that could be used as models?

Dewar: Alaska has actually done a really good job of managing their resources. They have sustainable halibut and sustainable salmon fisheries. I am sure there are others out there but [Alaska] has a really good long-term record of sustainable fishing.

SR: Do you know of anything with more potential—aquaculture for example?

Dewar: Aquaculture definitely, if done well, could help provide food to peoples' tables and help meet their protein needs. Aquaculture does have some problems. For example, there is a lot of debate about whether aquaculture fish are enough like wild fish that it is a good idea to enhance wild stocks with them; there are some dramatic differences between wild and aquaculture fish. There can also be disease and environmental problems where aquaculture is located. I know it can be done well; it is just a matter of making sure that there are no bad side effects.

SR: In what strategy do you hold the most faith?

Dewar: Certainly, for sustainable fishing, Individual Transferable Quotas have worked well because that gets around the problem of the tragedy of the commons. The fishermen are given stock in a fishery and what they can fish the next year depends on what they fish this year, so there is this long-term incentive for sustainable fishing. I do think that more economics and social sciences need to be brought to the table and that the politics need to be taken away so that it is based more on science and less on politics.

SR: How do you feel about ecolabeling?

Dewar: I think that it is a good tool to increase peoples' awareness. I do not think it is the only solution we are going to need. I think we are going to have to do a lot of other things as well as ecolabeling.

To verify ecolabeling, DNA barcoding confirmed the identity of thresher shark samples purchased at a San Diego fish market.

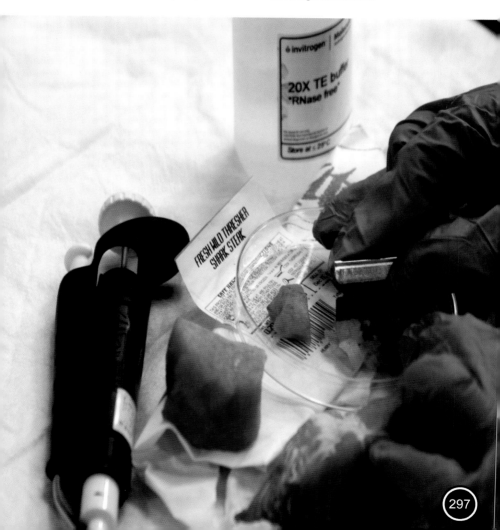

Jack Webster

We arrived at the commercial fishing docks along the harbor of San Diego Bay to meet Jack Webster we quickly found out that Mr. Webster is not your ordinary fishermen. The captain and owner of the *Millie G*, an albacore boat, seemed part professor, part entrepreneur and part environmentalist all in one. Clearly he is a leader in his trade as president of the American Albacore Fishing Association. As the floating deck rocked up and down we felt at sea with this sustainable fishing spokesperson. We realized that Mr. Webster truly lives by the principles of sustainable fishing, when we discovered that his fish is even sold at the ecofriendly Whole Foods Market.

Student Researcher (SR): Where do you see the overall population of fisheries in the future?

Jack Webster: All fisheries are different. An astute fisher once told me this: certain fish have an address and certain fish don't. Fish that do not have an address are difficult to identify, locate and catch. Fish that have an address live in a generally vicinity for their lifetime. Certain fisheries that actually target species that are born and raised around a certain area, can be highly susceptible to overfishing. Also offshore species are subject to overfishing, because certain countries subsidize, meaning they give money to their fishermen. These countries do this to help their fleets develop and get stronger and help employ people.

SR: What country is the most helpful in the sustainable fishing venture?

Webster: The United Kingdom (U.K.) and Norway are very committed to it. In fact that is where the Marine Stewardship Council started. They have taken the initiative to try and do something about the fishing problems that arise in their countries. Also the public at large, unlike here in the United States, has been taught about what went wrong and what can go right.

SR: Do you think that enough is being done to support sustainable fishing or sustainable fishermen?

Webster: To answer your question, yes. In certain parts of the world, especially in the U.K. and the European Union countries, the sustainable fish topic is a lot more prevalent in the media and the public at large. It has not taken off much here in the United States. It might just be complacency, but you are starting to see more and more of it here. So it is starting to get a foothold.

SR: How much do international affairs and politics play into this?

Webster: It comes in quite a bit because some nations have more resources then others when it comes to fishing. Politically it is becoming a hot button but there is only so much you can do. As American fishermen, the most highly regulated fishermen in the world, if there is a problem then you have to go and work it out with our fellow fishermen. Some of the other countries will never adhere to the same laws that we have. So if you do not like what they are doing then you do not have much of say in it. They are going to say that their fishermen are making money and employing people; they are going to keep doing it in the same way.

SR: Would you say that sustainable fishing is a very important political issue?

Webster: Yes, and I am sure there are people that are going to politically get on the bandwagon just for their own self-centered beliefs and self-centered aspirations. What I see is a way to create awareness. People are not going to simply become lemmings and jump off a cliff, but people will wonder what sustainable fishing is all about and then research and study it themselves to make their own decisions.

SR: What do you think is the best way to raise awareness?

Webster: Talk about it, if you are informed and you understand how it is, then you can't be afraid to talk about it and there are ways to look for it—web sites, and things like that. If you look up sustainable and Marine Stewardship Council and different fisheries that they have certified sustainable, then that is how you are going to find out about it.

Sustainable fishing practices must be taught to the next generation in order to preserve marine life.

Marine Protected Areas

"The ocean is ill and not doing as well as it could be, and I think it's our responsibility to help to manage that."

— Andrea Compton

The majority of the oceans are invisible to those who are destroying them. The lasting impact of overharvesting on land is much more apparent than the overexploitation of oceanic resources. A good example is the stark contrast between a cleared forest and a virgin wooded area. Because of the visible and extensive damage caused by clearing, the logging industry has been forced to leave regions of forest untouched. Only recently has the same idea been proposed for our oceans; the global community must work to salvage what life remains in them. (Arnett 2006)

The world's oceans were once teeming with fish and other marine creatures. Thousands of different species lived in this safe, aquatic refuge that covers 71% of Earth's surface. However, over time, humans have plundered the seas to a global state of alarm. Marine populations are becoming extinct, food sources are limited, and oxygen supplies are dwindling; all of this turmoil and pain caused by humans. We have watched as species such as tuna decrease in size and abundance.

The world faces an unprecedented period of environmental and ecological crises caused by post-industrial society's rapidly increasing population. It is only now dawning on us that the human impact on the environment may be irreversible. At this point, the best efforts of established individuals and organizations to correct the destruction may not return the environment to its original condition. Environmental awareness, conservation and restoration have become our primary tools in the battle we now face: eliminating the human footprint.

Given the need to restore and safeguard our environment, one solution that has been proposed for the world's oceans is establishing Marine Protected Areas (MPAs)—specific areas that regulate human activity so that marine ecosystems may replenish. These regulated areas present an interesting solution to the some of the ocean's environmental issues by not allowing humans access to selected ecological resources. Complemented by other solutions, MPAs can help to correct the harm humanity has inflicted upon the Earth.

The establishment of a single MPA can take years. Often, the affected resources are disputed by fisheries, tourists and conservationists. Some people would say that humans can meddle in nature because it is either inexhaustible or repairable. Thus, people often fear that the establishment of an MPA will hinder their access to the ocean's abundant natural resources. Such areas do hinder access, but we must create and sustain marine ecosystems within MPAs to allow the oceans to thrive. By doing so, we might help by allowing MPAs to serve as restorative foundations for nearby exploited environments. (Smith 1997)

To resolve user conflicts, MPAs must have zoning regulations. Humans depend upon the oceans for food and nature's supply is challenged by man's competition for access and volume. In their unrestricted scrabble, resources vanish at such a rapid rate that there is no chance for them to replenish. As competing fisheries seek endless profit, few notice the generally declining population trends before a threshold is passed. To meet and sustain the needs of industry and individuals alike, regulations are necessary so that oceanic life does not collapse and its resources are available to future generations. (Häder 1997)

Typically, MPA regulations are considerably varied. The most common restriction pertains to those who may or may not enter a specific area. For example, in a Uniform Multiple Use Area, the limitations on activities are consistent throughout the entire MPA. In Zoned Multiple Use MPAs, the MPA is split into different sections with varied regulations. Another key part of MPA regulations is the amount of harvesting permitted. No-take zones are fairly uncommon. Therefore, MPAs are required to specify whether or not commercial or recreational harvesting of fish, plant-life and other species is permitted. ("MPA Connections" 2005)

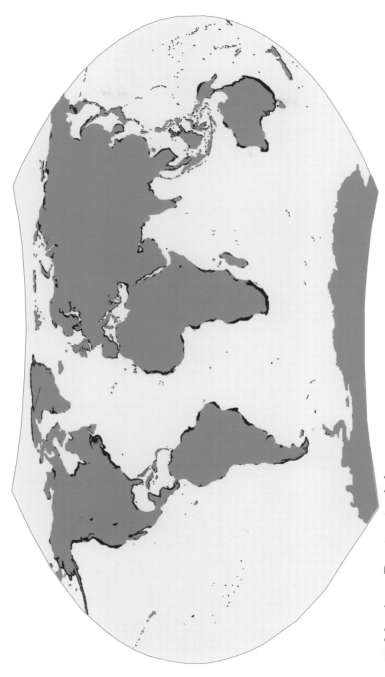

■ Marine Protected Areas

The performance and success of an MPA is difficult to measure; the effectiveness of an MPA is judged by its environmental results. This is done mostly through studying the proliferation of animal, algae and plant populations. Scientists report on species' numbers both before and after the designation of an MPA. These studies take years to produce results. Even then, there are so many variables within the ocean that the accuracy of these numbers may be questioned. This lack of conclusive answers can cause conflicts for those that are fighting to promote conservation. (Sobel and Dahlgren 1998)

For MPAs to succeed, nations that take advantage of the ocean's resources must not only participate in MPA networks, but realize how essential they are to long-term survival. Since long-term needs may conflict with their short-term desires, what may be beneficial in time is often sacrificed for faster, yet less significant progress. Unless governments change their policies, merely designating an MPA will never prove to be effective. This social or political aspect of MPA effectiveness is often more relevant to direct users. Frequently, the data are significantly more complicated than simply measuring changes in population size. To derive this overall value, it is important to analyze the use (both direct and indirect) and nonuse values of coastal waters. Direct use includes things like fisheries and oil extraction; indirect use is shoreline protection and nutrients in the ocean, such as oxygen; nonuse values are a creature's general worth and the overall health of the ocean. Because of the

MPAs, like the Mia J. Tegner State Conservation Area off Point Loma, help to protect local marine life that would otherwise be further exploited.

complexity of these values, the effectiveness of an MPA is often limited to theoretical and conceptual terms. (Sobel and Dahlgren 1998)

Social acceptance and public opinion greatly affect the production and effectiveness of an MPA. Without compliance from local users, an MPA will only prove to be nothing more than a series of legal documents. There is a huge risk that individual nations will plunder or exploit understocked resources. This is particularly common in developing nations, who often place the challenge of meeting the needs of their people over the implications of their actions. While developed nations made ecologically harmful mistakes during their industrialization, considering our newly found perspective on the fragile environment and limited natural resources developing nations cannot be allowed to do the same. (Smith 1997)

Another interesting case is that of China, a nation rapidly becoming an economic powerhouse. That alone gives it huge influence over international politics. In addition, its massive population means that a federal law regulating environmental degradation could affect the planet on a significant level. As of this writing, conservation is not one of China's highest priorities; the country believes it is entitled to the same latitude that England and the United States enjoyed during their industrialization. But during our period of industrialization, the U.S. population ranged from 23 to 100 million, compared to modern

China's 1.3 billion people. Even today, the 300 million U.S. citizens dwindle in comparison to the Chinese population. So, while our past errors made a lasting impression on the planet, they may be nothing in comparison to the impact China may have in the future. (Gage 2006) ("China" 2008) ("United States" 2008)

The United States has attempted to become involved in promoting conservation in China. Many U.S. organizations are working to establish an MPA in the South China Sea—an important project because of the vast resources found in those waters. China is renowned for its fishing, which puts it at risk of overfishing many species in this region. With an MPA in place, the population of species affected will have a chance to rebound and fishing can continue in the area. However, if their fishing industry is not regulated now, then it risks eliminating its product—fish. ("South China Sea" 2006)

In contrast, the United States is one of the leading nations in the development of MPA protocol. At a federal level, MPAs are controlled by the National Oceanic and Atmospheric Administration (NOAA), which is in charge of providing and researching information on environmental changes and is involved in environmental stewardship and advocacy projects throughout the nation. ("NOAA, AOML State of the Ocean" 2007)

Within NOAA, there is a subdivision entitled the National Marine Protected Areas Center, the group responsible for monitoring and maintaining federal MPAs. In 2000, Presidential Executive Order 13158 declared that the United States needed to develop a framework for the national system of MPAs. The National MPA Center was created and now oversees more than 1,500 U.S. marine managed areas (MMAs). The center was made responsible for developing an MPA inventory—the first step toward the goal of establishing an MPA network—that demands an understanding of levels of protection and how they are affecting the world's oceans. The inventory was finished in August 2006, and the center is now working on establishing a model for a functional network. ("NOAA" 2007)

California is developing a model for a network of small MPAs. The goal is to address the issue of protecting migrating creatures. By creating smaller MPAs, we will be allowing the fishing and tourism industries to continue to function while, at the same time, taking into consideration the constant shifting activity of pelagic species. California has currently installed more than 60 MPAs; however, this number is expected to grow over the next few years because of the West Coast Pilot Project, a study of the effectiveness of network MPAs. In this case, the California coastline will be used as a model for future networks in other regions. ("MPA Connections" 2005) (Häder 1997)

The effect of the West Coast Pilot Project on San Diego Bay remains unclear. With the bay's high toxic level, the pilot might not

The mouth of San Diego Bay is a part of the Mia J. Tegner State Marine Conservation Area, the largest kelp forest in California.

have a significant impact. But if San Diego Bay becomes a suitable habitat, populations of fish (including tuna, white seabass and wooly sculpin) could migrate from MPAs and return to their former habitats. Also, creatures in the bay could use the MPA and repopulate before returning to the bay's waters. (McArdle 2002)

Several MPAs are located off the San Diego coastline. These include the San Diego-Scripps State Marine Conservation Area and the La Jolla State Marine Conservation Area. The MPA with the greatest impact on San Diego Bay is the Mia J. Tegner State Conservation Area (honoring the late local marine researcher who did so much for the analysis and understanding of kelp forest ecology). Located near the mouth of the bay on the Cabrillo National Monument, the reserve was established in 1978 for the protection of the fragile, intertidal zone that harbors creatures like sea hares, octopi, hermit crabs and various fish and invertebrates. This MPA replenishes an environment damaged by ship activity and dredging operations in San Diego Bay that took place in the twentieth century. ("MPA Connections" 2005)

Local MPAs create a safe haven for native species previously forced to live in regions abused by humans. At the Tegner Conservation Area, a long-term study is being conducted on many different species, including the giant owl limpet, golden rockweed and boa kelp. (Becker 2006)

By using MPAs in local and state waters as a foundation for future plans, we can successfully implement these conservation tools. For example, though studies have suggested that regulating the capture of certain species is beneficial, we have learned that it is actually much more effective to simply prevent all harvesting in a small area. Also, we have realized that to make MPAs effective, it is vitally important to speak directly with individuals who use the region rather than just seeking action through local officials. By looking at issues that arise during the implementation of local MPAs, we can avoid the problems that may inhibit future plans.

Networking MPAs is not the only new idea being proposed. Another concept under debate is that of open-ocean protected areas. MPAs around the world provide direct protection to very limited areas of water; in fact MPAs, "cover around half a percent of the world's seas." (Norse and Crowder 2005)

By creating more MPAs in international, open waters, it would become easier to protect the millions of creatures that inhabit

Dana Point is an important MPA in Orange County.

the oceans. However, this presents the problem of enforcement. Since no nation would have the right to control these regions, they would be internationally governed. The current debate is over which nations could monitor and enforce the regulations for these areas. Considering the lengthy process involved with refining international policies, it will probably be some years before the plans come close to being a reality. (Norse and Crowder 2005)

While MPAs present one solution to our growing environmental and ecological crisis, there are still many issues associated with their implementation. Therefore, they must only be viewed as a single tool for combating one element of the greater issue mankind faces. The errors of man have led to the destruction of many species. Viable oceanic areas are shrinking to the point of nonexistence. However, people can fix these mistakes by using tools like MPAs to reverse the negative effects of human contact. Without the implementation of these protected regions, many of the ocean's most threatened ecosystems may be lost forever. ("NOAA, AOML State of the Ocean" 2007)

Andrea Compton

Andrea Compton has been the Chief of Natural Resource Science at the Cabrillo National Monument since 2003. As chief, she manages the resources of the park and conserves the natural and historic wildlife so that species may continue to prosper for future generations. Before coming to San Diego 13 years ago, she was a senior biologist for an environmental consulting firm in Portland, Oregon. Prior to that she worked at San Diego State University and Mesa College. Since her arrival at the National Park Service, Ms. Compton has been involved in a number of research projects at Cabrillo, including shorebird monitoring.

The day was clear and bright when we arrived at Cabrillo to meet with her. We conducted our interview near the area's tide pools, with a spectacular view of San Diego Bay serving as our backdrop.

Student Researcher (SR): What are some of your current projects at Cabrillo?

Andrea Compton: Here at the park I work with all the environmental resources and natural components, whether they be land-based or marine-based. So it's a diversity of projects. Everyday is a little different. It ranges from inventory projects, where we catalogue what animals and plants and other organisms are here, to monitoring projects where we look at long-term changes in the area. Some of our projects started in 1990, so we are looking at 18 years of data from intertidal monitoring.

Another long-term project that was started in 1995 is herpetological monitoring—studying reptiles and amphibians. I also conduct joint projects with other members of the conservation area here on the peninsula, which includes five different agencies. Then there's the Mediterranean Coast Network, which is three parks: the Channel Islands National Park, Santa Monica Mountains National Recreational Area and the Cabrillo National Monument. I also manage the GIS (geographic information system) program for the park, which involves managing data and developing and creating maps. And then a big part of it is managing budgets, writing proposals, developing ideas and concepts and pursuing the funding for those projects.

SR: How does the National Park Service collaborate with MPAs?

Compton: The National Park Service (NPS) has done quite a bit of work with MPAs thus far, especially at the Channel Islands National Park. Here at the Cabrillo, the Mia J. Tegner State Marine Conservation Area is a strip of land from the mean high-tide level to 45 meters offshore, and it is a state designation that provides some protection within this rocky intertidal zone. It does allow for the commercial and recreational harvest of marine plants and fin fish. Federal regulations actually have a much stronger influence on and restrict further what can be taken from the area. When the NPS slowly gained land and property down to the water's edge, an agreement was set up with the Navy that the NPS would have administrative jurisdiction over those waters. It is that administrative jurisdiction that gives us the authority to manage this rocky intertidal zone.

We have a long-term monitoring program in place where we study the species that are present there. We have a very active tidepool educator's program; volunteers are present at the rocky intertidal zone and they talk with the public to make sure that no plants or animals are taken. We also have set aside one portion of the area where the public is not allowed so the area may recover. Most of our management thus far is land-based [rather than aquatic] within the rocky intertidal zone.

The federal protection started before the park even existed. Having this protected area in place, whether it is through state or through federal conservation, allows us to really improve visitor experiences and maintain this habitat that would not be possible otherwise. Protecting our habitat is just a critical part of being able to provide the high quality experience that we offer. We also have researchers come and visit since the area is an increasingly interesting place for them.

SR: What type of ecological impact do MPAs have?

Compton: MPAs tend to become little underwater islands. You can imagine that if you have a series of them up and down the coast they become a chain of islands, which in turn become a refuge for fish. Fish can grow larger and then move out to other areas from there. So I think MPAs are very important for our local marine populations to survive. And you think about population growth in San Diego; growth is occurring and will continue to occur. I think that additional use of the ocean will also grow—through diving and through additional fishing—and all of those things can be supported if MPAs are set up effectively. For example, one researcher that I talked to recently said that the Point Loma kelp beds have been really barren. Well, the Point Loma kelp bed is one of the longest kelp beds on the coast. So, while this impact is unfortunate, it is an amazing opportunity to add some protection by establishing an MPA that would help to preserve it. The state conservation area only goes out 45 meters, and even though our federal administration jurisdiction goes out 275 meters, discrepancies occur. For example, crab and lobster fishing occurs within those 275 meters even though it's against federal regulations. The advantage of the MPA is that it specifically covers fishing regulations, so fishermen are more likely to recognize the limitations that occur right along the park edge.

SR: Why are MPAs an important part of marine protection, especially in San Diego?

Compton: I think MPAs are going to be an important part of helping San Diego to preserve our ocean resources far into the future. The ocean is ill and not doing as well as it could be, and I think it's our responsibility to help to manage that. I think MPAs have and will have a strong ecological and economical impact for San Diego. We have some high-quality intertidal habitats, established largely through the efforts of the NPS, which can be extended to the offshore habitats as well. And through our providing MPAs, I think that will allow the recovery of the ocean, the recovery of our offshore habitats, and provide more opportunities for the residents of San Diego.

THE BARNACLE

Long ago when I was floating
Adrift along the sea
I found a rock, so high on shore
That was my home-to-be.

I latched my head and lo behold
I soon began to grow.
A Barnacle became of me!
My fate, I did not know.

Every day I'd wait the tide
Oh up and down it went
And every time the cycle passed
In feast, my time was spent.

When high tide comes the famine ends!
But now the plankton's bare?
Of food, and life, and times long past:
Now famine, drought, and scare.

The tides have changed and soon I see
The humans living near.
They're coming close! And soon a CRUNCH!
In death, there is no fear.

— Megan Morikawa

Carl Hubbs

"He wanted to be the best biologist in the world."

— Clark Hubbs

Carl Leavitt Hubbs (1894–1979), a leading marine biologist of his time, invested much of his life to ocean conservation and helped to change human views on the environment. Though Hubbs developed interests in many areas of marine biology, his passion was ichthyology, the study of fish. He contributed greatly to that field, saving a subspecies of killifish from extinction and amassing a marine collection of millions of specimens. A major portion of his contribution to marine science took place during his appointment at Scripps Institution of Oceanography. Not only did he participate in many research projects, but he continually added to the Scripps' library collection.

This chapter covers only a small segment of Hubbs' life, but it is clear that he was a man of great accomplishments and dedication. However, centuries before him, there were others who contributed to the study and knowledge of nature and the oceans so that man could live in better harmony with the environment. Today's conservation and restoration efforts, such as those described in this book, are what Hubbs and earlier scientists had worked towards.

Hubbs was born in Williams, Arizona on October 18, 1894 but spent part of his childhood in San Diego before moving to Los Angeles, where he graduated from high school. At Stanford University his principal professor and mentor was the famous biologist David Starr Jordan. Jordan took a great liking to the young and eager undergraduate, describing him as,

> "… the ablest student I have had for the past thirty years …. There is no one now doing systematic work on fishes that has as keen an insight, or as accurate a mind as Hubbs, and he is tremendously industrious."

Hubbs obtained a bachelor's degree in zoology under Jordan in 1916. (Campbell 1979)

While at Stanford, Hubbs was assigned the task of curating the university's extensive fish collection, which led to his lifelong interest in ichthyology. He continued to pursue this passion in his graduate studies, receiving his master's from Stanford in 1917. During this time, he met the vivacious Laura Clark, a mathematician who would soon become a major part of Hubbs' life. They were married on June 15, 1918 and entered into a lifelong partnership. (Rosenblatt 2003) ("Historical Perspectives …" 1963)

Hubbs went on to become the curator of zoology at the University of Michigan's Museum of Zoology. At Michigan, he earned his Ph.D., and remained with the university from 1920 to 1944. (Rosenblatt 2003) ("Historical Perspectives …" 1963)

Over the years, Hubbs amassed a collection of fish that is still today one of the largest in the world, with more than two million specimens. Though mainly consisting of North American freshwater species, other fish from the U.S. West Coast, Japan and Indonesia were added. (Rosenblatt 2003)

In 1930, Hubbs became the director of the Institute of Fisheries Research, an organization formed by the joint effort of the University of Michigan and the Michigan Department of Conservation. He was responsible for overseeing and aiding the development of restoration techniques for aquatic habitats such as streams and lakes. From this research, Hubbs and his colleague, Karl Lagler, published a book entitled, "Guide to the Fishes of the Great Lakes and Tributary Waters," which included extensive descriptions of collection methods, data recording and preservation techniques. ("Carl L. Hubbs …" 1973) (Rosenblatt 2003)

Before joining Scripps, Hubbs had an association with the research facility; he coauthored several papers with the aquarium's curator. He had been offered a position several times by Harald Sverdrup, the institution director, and in 1944, Hubbs decided to finally accept. Upon arrival in La Jolla, he donated his personal library, reprint collection and papers that enriched the Scripps' library, already an extensive oceanographic resource. He would stay at Scripps until his death in 1979. Today at Scripps, Hubbs is remembered and honored for his contributions to education and conservation by a marine biology building in his name. (Rosenblatt 2003) (Frieman 1988)

Laura Hubbs took an active role in her husband's work. She organized his notes and files, helped to maintain aquarium tanks, corresponded with his massive list of contacts, and proofread his manuscripts. As Hubbs' work mostly consisted of field trips, she performed math calculations needed for projects, helped with the collection of specimens, and acted as a fellow notetaker. She also coauthored 19 of Hubbs' 787 published papers. At home she was a warm, pleasant hostess, mother of three, and caretaker of their household. Essentially, Laura Hubbs made it possible for her husband to succeed through her support and assistance, thus sharing a part in everything he accomplished. ("Laura Clark Hubbs …" 1988) (Cox 1981)

Hubbs' contributions to the field of ichthyology included the general taxonomy of fishes. He discovered many different species on his travels, which he brought back to Scripps. Measuring these species was an important task in which both husband and wife were involved; together they set the standard for their generation. (Rosenblatt 2003)

Hubbs is also acclaimed for his conservation of endangered species. Soon after arriving at Scripps, Hubbs' interest was sparked by the gray whale that was making a slow recovery from overhunting. He and Laura Hubbs would make annual visits to Baja California, Mexico to survey the whale's population. Hubbs demonstrated his negotiation skills by persuading the Mexican government to make Scammon's Lagoon, a large gray whale breeding ground, a

Hubbs Hall was built at Scripps Institution of Oceanography to commemorate the life and work of one of San Diego's most distinguished biologists.

protected environment. Hubbs also gathered a group of volunteers and students to help with an annual gray whale census during the animals' migration offshore from Scripps. (Rosenblatt 2003)

Two other marine mammals that concerned Hubbs were the Guadalupe fur seal (*Arctocephalus townsendi*) and the northern elephant seal (*Mirounga angustirostris*). The fur seal was thought to be extinct until a rare sighting on Guadalupe Island (off the Baja California peninsula) proved otherwise. Hubbs traveled to the island and, after a long search, he found a small colony of 14 fur seals on a secluded shore, and later some elephant seals. His observations and research of both animals' behavior, breeding and population numbers were later published in papers, and his' efforts on their behalf led to seals' official protection by the Mexican government. (Wisner 1985)

During his lifetime, Hubbs held many prestigious positions. He acted as review editor for the *American Naturalist* from 1941 to 1947. In 1951, he was elected president of the Society of Systematic Zoology. He was vice president of the Society for the Study of Evolution in 1953 and again in 1955. Later, in 1964, he held the same title with the American Society of Naturalists.

Hubbs also was involved in the Nature Conservancy, the Torrey Pines Association, the Desert Fishes Council and the American Society of Ichthyologists and Herpetologists. He acted as that society's secretary from 1928 to1931, its editor from 1930 to 1937, and its president from 1934 to 1935. He was instrumental in not only the society's success but the success of *Copeia*, which is now a major international zoological journal. An edition of *Copeia* honored Hubbs by dedicating an issue to him on his eightieth birthday. (Horn 1976) ("Carl L. Hubbs ..." 1973) (Wisner 1985)

Throughout his life, Hubbs was acknowledged in numerous ways: he received the Henry Russel Award (1929–1930) for his, "distinguished achievements in the field of scholarly research"; elected Fellow of the California Academy of Sciences, along with a Guggenheim Fellowship in 1933; elected to the National Academy of Sciences in 1952; honored with the Joseph Leidy Award and Medal from the Academy of Natural Sciences of Philadelphia in 1954; the Fellows Medal of the California Academy of Sciences in 1956; and the Gold Medal for Conservation from the San Diego Zoological Society in 1970. Hubbs had been appointed trustee of this society and was involved in its conservation efforts for 19 years—especially its zoo and wild animal park.

In 1973 Hubbs was the recipient of the American Fisheries Award for Excellence, and the San Diego Zoological Society and the San Diego Society of Natural History named him Scientist of the Year in 1974. That year he also received the Man of the Year award from the American Cetacean Society for his efforts on behalf of the gray whale and his, "unfailing interest in and aid to questioning students."

In the words of Dr. Michael H. Horn, who had been personally counseled by Hubbs after the publication of one of Horn's papers, he was, in many ways, "the foremost figure in North American ichthyology in the middle years of the twentieth century." ("Carl L. Hubbs … 1973) ("News Release" 1974) ("Carl L. Hubbs, Famed Marine Biologist, Dies" 1979) (Horn 1976) ("Historical Perspectives" 1963) (Campbell 1979)

Among the international awards Hubbs received was the Shinkishi Hatai Medal of the Japan Science Association in 1971, "for the most remarkable contribution to marine biology in the Pacific." Hubbs was also proud of the fact that he was an honorary member of the Japanese Ichthyological Society. Two other foreign societies that granted Hubbs affiliation were the French Society of Ichthyology and the Linnean Society of London.

All throughout these exciting years, Hubbs made frequent collecting trips not only in the United States but around the world to find and study new fish. (Campbell 1979) (Wisner 1985) (Rosenblatt 2003) ("Carl Leavitt Hubbs Biography" 2007) However, the highest honor that can be bestowed on a biologist is his or her name in the genus of a new species. This recognition Hubbs received numerous times. A bird, whale, two mollusks, one crab, three cave arthropods, two insects, three algae, one lichen and five fish have all been named for him; each bearing the designation *hubbsi*. ("Carl L. Hubbs …" 1973)

Carl Hubbs died from cancer on June 30, 1979 and his ashes were spread along the gray whale migration path of the ocean he had loved. Hubbs was survived by three children, all inspired by their parents' passion for biology. Dr. Clark Hubbs was a renowned ichthyologist at the University of Texas at Austin until his death in February 2008, Earl L. Hubbs is a biology teacher, and Frances married one of her father's students, Robert Rush Miller. Extended family at the time of Laura Hubbs' death in 1989 included 13 grandchildren and one great grandchild. (Campbell 1979) ("Carl L. Hubbs, Famed Marine Biologist, Dies" 1979) (Wisner, 1985) ("Laura Clark Hubbs …" 1988)

The impact Carl Hubbs made during his life has left a lasting impression not only in San Diego, but throughout the world. His passion and dedication expanded our knowledge of nature and changed the way many people viewed the environment. His life-long efforts to incorporate science and conservation is exactly what is needed to keep the healthy balance between man and nature, and save endangered species. It requires a collective understanding around the world, a far greater issue than that of San Diego Bay. Fundamentally, we need more Carl Hubbs in this region and in this world.

Story of the Killdeer

(Charadrius vociferus)

While we were chasing the conservation stories of many of San Diego's most treasured coastal wildlife, we found one that decided to make its home at our school. Over our two-week spring break, a pair of killdeer (*Charadrius vociferus*) settled in the then-empty parking lot of High Tech High in Point Loma. Finding the location peaceful, the killdeer built a nest on the ground by making a small depression in the bark of an island divider in the parking lot.

Fortunately, several days before school began, the director of the IT department notified our biotechnology teacher, Dr. Jay Vavra, about our school's newest guest. Though spring break was not yet over, Dr. Vavra came to school several days early in order to set up a barricade, which was composed of a plastic mesh fence and six metal stakes to hold it in place. Four cones blocked off the adjacent parking spaces surrounding the killdeer's nesting site. While admiring his handiwork, Dr. Vavra noticed that the killdeer had been separated from its nest and would not return, leaving the eggs unattended. In order to simplify access to its nesting site, a section of the mesh was lifted so the killdeer could walk in through the bottom. Almost immediately, the killdeer returned to its cluster of tantalizing, speckled eggs.

When school resumed, the barricade was effective in isolating the nest, as proved by our ability to watch from a safe distance as the killdeer incubated its eggs. When we approached, the sitting bird burst out in the familiar *kill-deee* call to warn other killdeer of our intrusion. After three weeks of incubation, the killdeer eggs hatched.

The killdeer chicks quickly hid under their mother as members of the High Tech High community attempted to document this strange nesting presence in the parking lot.

Wild animals often must adapt to urban environments. Killdeer have been among the more successful species in this area.

As soon as the chicks immerged from their eggs, they were off and running about the parking lot. This is a classic example of precocial development, a phenomenon where offspring are already reasonably self-sufficient immediately after birth or hatching. Due to their mobility in such a dangerous setting, it was a concern that some of chicks would survive the first week. However, nearing the end of that time, we were relieved to learn that all four chicks had survived this early stage of their lives.

With the chicks' development well underway, their parents' worries seemed to only increase. Like other plovers, when an intruder approaches a killdeer chick, one of the mature killdeer will attempt to lure the invader away. In a display of ingenious evolutionary development, the parent killdeer exposes its rust-colored tail feathers as a luring tactic. The adult killdeer then feigns a broken wing and hobbles away from its young, distracting the potential predator.

As it turned out, the chicks faced far greater danger than the threat of predators in the dangerous terrain of High Tech High's parking lot. During the course of their explorations, they would scurry about in search of food. On two separate occasions, a few particularly venturesome chicks fell through the bars of a nearby drainage grill and were essentially trapped in a subterranean concrete prison. Luckily for them, a group of students and faculty were able to lift the grating and free the chicks.

As soon as the killdeer chicks were able to fly, they fled the perils of the High Tech High parking lot and vanished. Unlike losing a pet, the departure of the killdeer family was greeted more with feelings of satisfaction and relief than sorrow and longing. We realized that though this was only a few members of one species that had been displaced by urban development, we were doing our part to set the example to the rest of the High Tech High community and beyond.

STEINBECK, RICKETTS, AND HOLISM IN SAN DIEGO BAY

On March 11, 1940, a small fishing boat called the *Western Flyer* departed from Monterey Bay, about 600 kilometers north of San Diego Bay. Among its crew was John Steinbeck, one of the most celebrated authors in American literature. Steinbeck and his close friend, Ed Ricketts, a respected marine biologist, organized the six-week voyage to the Sea of Cortez and back. The purpose of the trip was to collect and catalogue intertidal specimens, and it was extremely successful in this goal: more than 500 organisms were identified, about 50 of them being new species. This expedition also inspired some of Steinbeck's later works, including "The Pearl." The most valuable of the rewards, however, was far deeper: the life-altering sea journey profoundly affected the philosophies of both Steinbeck and Ricketts.

While exploring the Sea of Cortez, the men were completely immersed in the largely unknown region, spending their time studying the native life and discussing philosophy. Their conversations, which can be found in the published account of the voyage entitled "The Log from the Sea of Cortez," concerned the nature of life, man's role in the world, and the unending quest for knowledge and understanding. The two men arrived at the concept of holism—that all life is connected, and that the perceived divisions between one being and another are purely illusory. (Steinbeck, Ricketts, and Astro 1995)

Holism is at first difficult for many to understand and accept. This is especially true for modern Westerners, who do not often take time to examine and appreciate nature, and have come from a culture that traditionally does not emphasize life's unity. However, holistic themes are revealed through the study of both biology and philosophy. There is no question that diverse species interact with one another to survive, and for the survival of their respective ecosystems. Because these interactions are often extremely subtle, many of them remain unknown to us, our current methods being inadequate to document such intricate complexities. Food chains, crudely simplistic linear illustrations of organisms' dietary habits, are now often replaced by food webs, which demonstrate the fact that relationships between organisms are often much more elaborate and delicate. Perhaps it was John Muir, celebrated naturalist and founder of the Sierra Club, who best described the inevitable interrelatedness of life, "When we try to pick out anything by itself, we find it hitched to everything else in the Universe."

There are even times when organisms are so cooperative that biologists view them as a single entity, called a "superorganism." Ants, bees and termites familiarly demonstrate this phenomenon: each individual plays a role in the survival of the colony, which seems to have a mind and consciousness of its own that continues on, even after its members pass. Superorganisms also have unexpectedly vast potentials. An entire colony of ants working together towards a common goal can solve problems that seem beyond their intelligence, and accomplish feats that seem beyond their strength. The colony is much more than a large number of individual ants, much like the human body is more than a large number of individual cells. Thus, where life is concerned, the whole is greater than the sum of its parts, a fact which compels us to consider the possibility that there exists in life an underlying spirit that transcends the shallow world of our perception.

Some biologists have suggested that the entire biosphere of Earth is simply a superorganism in itself, and that each living thing is therefore a part of the great colony of life. And indeed, Earth qualifies fittingly for the title of superorganism, shifting and adapting to its own natural changes and outliving the beings that comprise it. William Emerson Ritter, the biologist whose work influenced both Steinbeck and Ricketts, once wrote, "In all parts of nature itself as one gigantic whole, wholes are so related to their parts that not only does the existence of the whole depend upon the orderly cooperation and interdependence of the parts, but the whole exercises a measure of determinative control over its parts." (Ritter 1919) (Steinbeck, Ricketts, and Astro 1995) Holism can be interpreted from a variety of perspectives. Biologists may call it large-scale symbiosis, followers of a number of religions may call it a collective soul, and others may see it as a metaphor for the communal environment of the world. (Steinbeck, Ricketts, and Astro 1995)

Steinbeck's holism is related to the concept of "biophilia," proposed by renowned biologist, Dr. Edward O. Wilson. In his book, "Biophilia," Dr. Wilson suggests that life, specifically mankind, has an inherent affinity for, and bond with, all other forms of life. This accounts for our love of domestic pets and decorating our homes with attractive plants. Biophilia represents the natural love of life and living systems, a result of the intrinsic connection that exists between all life forms. (Wilson 1986; 2008)

Returning to superorganisms, mankind could be viewed as a member of this class as well. Like ants, humans work cooperatively with one another towards achievements that would be impossible if attempted alone. Interestingly, the difference between mankind and other superorganisms is that humans do not always have common goals, and may exhibit uncooperative and even self-destructive behavior patterns. Unlike all other organisms, those which Ricketts called, "the good, kind, sane little animals," mankind often seems content to destroy members of its own species, and inflict irreversible damage on its environment—our own global, superorganismal home. Perhaps this is why Steinbeck once referred to man as, "a two-legged paradox." Since this is the case, there is clearly a serious imbalance in the natural order of the world. As an animal, mankind has existed in relative harmony with Earth until only the past two centuries. Humanity's current behavior, however, directly conflicts with the foundations of holism and biophilia. (Steinbeck, Ricketts, and Astro 1995) (Ricketts and Steinbeck 1992)

Steinbeck wrote that, "Every new eye applied to the peephole which looks out at the world may fish in some new beauty and some new pattern, and the world of the human mind must be enriched by such fishing." (Ricketts and Steinbeck 1992) If all life is essentially part of one whole, then each time the last member of a species perishes, however distant or obscure, a part of the world's biodiversity and a part of every human being is irretrievably lost. And who knows what living gems we are losing every day? One species might have changed our entire understanding of biology, another may have held the secret to a seemingly incurable human disease. Because of the misdeeds of our ancestors, it is now impossible for us to behold a dodo or a Carolina parakeet. And as extinctions continue apace, every day we reduce the world's biodiversity for our descendants, who will never behold a Baiji River dolphin or a Sweetwater River rainbow trout. Humanity's destructive, parasitic actions must be reversed if we wish to claim compassion and respect for Earth, and subsequently for ourselves.

And yet, it is not always this simple. This book has illustrated a number of complex issues with no clear solutions. What should be done, for instance, when imperiled gull-billed terns prey upon endangered California least terns? Is it better to allow the population of one endangered species to diminish, or to deny another a natural food source? Another example is found with rock pigeons and European starlings, bird species introduced from Europe that thrive in urban environments, and often displace struggling native animals. Is it better in cases like this to begin organized population control through selected killing of these animals, who are themselves guilty of no crime, or to allow them to put native species at risk? Dilemmas like this raise questions that echo throughout conservationism: when should we simply leave nature alone to avoid causing further damage, and when is it our responsibility to step in and take action? These are the types of difficult questions that often face environmental stewards.

Perhaps the greatest and most important of these complicated matters is how best to balance environmental preservation with human interests. The destructive acts that threaten Earth today are

The aggressive and adaptable American crow often displaces other native species. Here, one prepares to attack a perching red-tailed hawk.

often committed without malice. Regardless, when each human on Earth is somewhat environmentally irresponsible, the collective result is a devastating global problem. The developed countries of the world, those with the most money and resources, must step forward and act more wisely regarding ecological issues, working to conserve habitats and biodiversity hotspots in poorer countries, which often have more pressing concerns than establishing protected areas. Sadly, during the final stages of this book, the United States was only a party to the United Nations' Convention of Biological Diversity and not to the subsequent Cartagena Protocol on Biosafety.

Though the ecological state of Earth is certainly not the only topic that deserves our concern, it is one to which the entire world needs to give its immediate attention. And while it is certain that biodiversity has to be preserved and environmental degradation stopped, the ideal methods for meeting these goals are less obvious.

When surrounded by tragic circumstances it seems nearly impossible to resist despair, but it is during such times when hope is stronger than ever. It is important to remember that if Earth is to be considered a superorganism, it will fight for its own survival. As the stories in this book have demonstrated, humans must be involved with this process, building an understanding of ecological issues and conscientiously considering how our actions will affect the entire biosphere. It is fortunate that there are restoration and recovery efforts, such as those that have been outlined in the preceding pages, that endeavor to realign humanity with its holistic baseline. Holism, as discovered by Steinbeck and Ricketts, is an enduring philosophy that, as a part of the global whole, places responsibility on every individual to adopt sustainable ways and take positive, compassionate action towards the preservation of Earth.

Bibliography

Biodiversity

"California Floristic Province." Conservation International. 26 February 2007. <http://www.biodiversityhotspots.org/xp/Hotspots/california_floristic/>.

City of San Diego Vernal Pool Inventory. San Diego, California: Planning Department, Multiple Species Conservation Program, 2004.

Ezcurra, E. *Personal Interview*. San Diego Natural History Museum. 3 April 2007.

Groombridge, B., and M.D. Jenkins. *World Atlas of Biodiversity: Earth's Living Resources in the 21st Century*. Berkeley: University of California Press, 2002.

"National Invasive Species Information Center." U.S. Department of Agriculture. 26 February 2007 <http://www.invasivespeciesinfo.gov/>.

"Protected Areas." Convention on Biological Diversity. 21 March 2007. <http://www.biodiv.org/programmes/cross-cutting/protected/default.asp>.

"SDAS Conservation Projects." San Diego Audubon Society. 28 February 2007. <http://www.sandiegoaudubon.org/conservation.htm>.

Stein, B.A., L.S. Kutner, and J.S. Adams, eds. *Precious Heritage: The Status of Biodiversity in the United States*. New York: Oxford University Press, 2000.

Unitt, P. *San Diego County Bird Atlas*. San Diego: Ibis Publishing, 2004.

"Wilson Life and Work." E.O. Wilson Biodiversity Foundation. 20 May 2008. <http://www.eowilson.org/index.php?option=com_content&task=view&id=43&Itemid=69>.

Wetlands

California Fish and Game Code. §2785 (g), Sacramento, CA. 20 May 2008. <http://law.justia.com/california/codes/fgc/2785-2799.6.html>.

Coastal Environments. *Los Penasquitos Lagoon Baseline Biological Study Habitat and Bird Surveys*. Carlsbad, CA: Los Penasquitos Lagoon Foundation, 2003[1].

——. *Potential Enhancement Projects for Los Penasquitos Lagoon*. Carlsbad, CA: Los Penasquitos Lagoon Foundation, 2003[2].

Cylinder, P.D., et. al. *Wetlands Regulation, A Complete Guide to Federal and California Programs*. Point Arena, CA: Solano Press, 1995.

Goldstein, J.H. 1996. Whose land is it anyway?: Private property rights and the Endangered Species Act. *Choices, the Magazine of Food, Farm, and Resource Issues*, 4–8.

Greer, K.A., and D. Stow. 2003. Vegetation Type Conversion in Los Penasquitos Lagoon, California: An Examination of the Role of Watershed Urbanization. *Environmental Management*, 31(4):489–503.

Greeson, P.E., J.R. Clark, and J.E. Clark, eds. *Wetland Functions and Values, The State of Our Understanding*. Minneapolis: American Water Resources Association, 1979.

Kusler, J. *Wetland Creation and Restoration*. New York: McGraw-Hill Book Company, 1990.

Pacific Estuarine Research Laboratory (PERL). *A Manual For Assessing Restored and Natural Coastal Wetlands*. La Jolla: California Sea Grant College Program, 1990.

Patten, B.C., S.E. Joergensen, and H. Dumont, eds. *Wetlands and Shallow Continental Water Bodies*. The Hague, Netherlands: SPB Academic Publishing, 1990.

Peugh, J. *Personal Interview*. 12 December 2007.

Spink, F., T. Black, and D. Porter. *Wetlands: Mitigating and Regulating Development Impacts*. Washington, D.C.: Urban Land Institute, 1991.

Shifting Baselines

"A Worldwide Trend: MPAs." ShiftingBaselines.org. 28 February 2007. <http://www.shiftingbaselines.org/mpas/psa.php>.

Jackson, J. *Personal Interview*. 14 March 2007.

McKeever, M. *A Short History of San Diego*. San Francisco: Heyday Books, 1985.

Olson, R. "Shifting Baselines: Slow-Motion Disaster in the Sea." Action Bioscience. 2 March 2007. <http://www.actionbioscience.org/environment/olson.html>.

Pauly, D. 1995. Anecdotes and the Shifting Baseline Syndrome of Fisheries. *Trends in Ecology and Evolution,* 10:430.

Shelvocke, G. *A Voyage Around the World by Way of the Great South Sea*. London: J. Senex, 1726.

"The Team." ShiftingBaselines.org. 25 March 2007. <http://www.shiftingbaselines.org/team/index.html>.

DNA Barcoding

Folmer, O., et al. 1994. DNA Primers for Amplification of Mitochondrial Cytochrome c Oxidase Subunit I from Diverse Metazoan Invertebrates. *Mol. Mar. Biol. Biotechnol*, 3:294–99.

Lorenz J.G., et al. 2005. The Problems and Promise of DNA Barcodes for Species Diagnosis of Primate Biomaterials. *Phil. Trans. Roy. Soc. London*. 360(1462):1869–77.

Palumbi, S.R., and C.S. Baker. 1994. Contrasting Population Structure from Nuclear Intron Sequences and mtDNA of Humpback Whales. *Mol. Biol. Evol,* 11:426–35.

Vavra, J., and O. Ryder. "A class action: DNA barcoding, student scientists and the bushmeat crisis." *ZOONOOZ*, Oct. 2006.

Abalone

High Tech High. *San Diego Bay: A Story of Exploitation and Restoration*. pp. 36–43. La Jolla: California Sea Grant College Program, 2007.

Leet, W.S., et al., eds. *California's Living Marine Resources: A Status Report*. Sacramento, CA. California Department of Fish and Game, 2001.

Morris, R.H., D.P. Abbott, and E.C. Haderlie. *Intertidal Invertebrates of California*. Stanford: Stanford University Press, 1980.

Rugh, S. *Personal Interview*. 2007.

Green Sea Turtle

Dutton, P. *Personal Interview*. 16 May 2007.

"Green turtle" 14 August 2008. <http://en.wikipedia.org/wiki/Green_Sea_Turtle>.

Hawxhurst, J.C. *Turtles and Tortoises*. pp. 61–3. San Diego: Lucent Books, 2001.

"Marine Turtles." NOAA Fisheries. 1 April 2007 <http://www.nmfs.noaa.gov/pr/species/turtles/>.

Stinson, M. *Personal Interview*. 16 May 2007.

"San Diego Bay Integrated Natural Resources Management Plan." U.S. Department of the Navy, Southwest Division (USDoN, SWDIV), San Diego, CA. September 2000.

California Brown Pelican

Anderson, D.W., F. Gress, and K.F. Mais. *Brown Pelicans: Influence of Food Supply on Reproduction*. Monterey: Blackwell Publishing, 1982.

Beacham, W., F. Castronova, and S. Sessine. *Beacham's Guide to the Endangered Species of North America*. Vol. 1. New York: Gale Group, 2001.

Briggs, K.T., et al. 1981. Brown Pelicans in Southern California: Habitat Use and Environmental Fluctuations. *The Condor* 83:1–15.

Burkett, E.E., R.J. Logsdon, and K.M.Fien. 2007. "Report of the California Fish and Game Commission: Status Review of California Brown Pelican (*Pelicanus occidentalis californicus*) in California." Nongame Wildlife Program Report 2007-04. 26 pp. + app. California Department of Fish and Game, Wildlife Branch, Sacramento, CA.

Pearson, T.G. *Birds of America*. Garden City, NY: Garden City Books, 1936.

Peterson, R.T. *A Field Guide to the Birds*. Boston: Houghton Mifflin Company, 1934.

Small, A. *Birds of California*. New York: Winchester Press, 1974.

Steinbeck, J., E. Ricketts, and R. Astro (Introduction). *The Log from the Sea of Cortez*. (Reprint edition). New York: Penguin Books, 1995.

Unitt, P. "Brown Pelican." *San Diego County Bird Atlas*. San Diego: Ibis Publishing, 2004.

American White Pelican

Anderson, D.W., and D.T. King. 2005. Introduction: Biology and Conservation of the American White Pelican. *Waterbirds*. 28(1):1–8.

Audubon, J.J. *Birds of America*. USA: Publisher unknown, 1842.

Carson, R. *Silent Spring*. Boston: Mariner Books, 2002.

Murphy, E.C. 2005. Biology and Conservation of the American White Pelican: Current Status and Future Challenges. *Waterbirds*. 28(1):107–12.

Peterson, R.T. *A Field Guide to the Birds*. Boston: Houghton Mifflin Company, 1934.

Rocke, T., et al. 2005. The Impact of Disease in the American White Pelican in North America. Wildlife Damage Management, Internet Center for USDA National Wildlife Research Center – Staff Publications. pp. 87–94.

Shuford, W.D. 2005. Historic and Current Status of the American White Pelican Breeding in California. *Waterbirds*. 28(1):35–47.

Small, A. *Birds of California*. New York: Winchester Press, 1974.

Udvardy, M.D.F., and S. Rayfield. *The Audubon Society Field Guide to North American Birds—Western Region*. New York: Alfred A. Knopf, Inc. 1977.

California Least Tern

Burr, T.A., and T. Conkle. 2003. Management of California Least Terns and Western Snowy Plovers on Naval Base Coronado. Presented at the Navy Natural Resources Conference, San Diego.

"California Least Tern IBA/Habitat and Education Project." Audubon Society: San Diego Chapter. 6 March 2007 <http://www.sandiegoaudubon.org/conservation. htm>.

"Least Tern *Sterna antillarum*." U.S. Geological Survey. 12 December 2006. <http:// www.mbr-pwrc.usgs.gov/id/framlst/Idtips/h0740id.html>.

Marschalek, D.A. 2007. "California Least Tern Breeding Survey—2006 Season." Nongame Wildlife Branch Report 2007-01. California Department of Fish and Game, Wildlife Branch, Sacramento, CA.

Patton, R. Terns of San Diego County. 29 March 2007. Presented at the San Diego Audubon Society meeting, Tecolote Canyon Nature Center.

"SDAS Conservation Projects." Audubon Society: San Diego Chapter. 6 March 2007. <http://www.sandiegoaudubon.org/conservation.htm>.

Unitt, P. "Least Tern." *San Diego County Bird Atlas*. San Diego: Ibis Publishing, 2004.

U.S. Fish and Wildlife Service. 1918. *Migratory Bird Treat Act of 1918*. Digest of Federal Resource Laws of Interest to the U.S. Fish and Wildlife Service.

Elegant Tern

Baughman, M. *National Geographic Reference Atlas to the Birds of North America*. Des Moines, IA: National Geographic Society, 2003.

Boswall, J. and M. Barrett. 1978. Notes on the Breeding Birds of Isla Raza, Baja California. *Western Birds,* 9:93–108.

Collins, C.T. 2006. Banding Studies of Elegant Terns in Southern California. *North American Bird Bander,* 31:17–22.

"Elegant Tern." National Audubon Society. 26 February 2007. <http://web1.audubon.org/science/species/watchlist/index.php>.

"Elegant Tern." 2003. Cornell Lab of Ornithology. <http://www.birds.cornell.edu/AllAboutBirds/BirdGuide/Elegant_Tern.html>.

Ezcurra, E. *Personal Interview*. San Diego Natural History Museum. 3 April 2007.

Patton, R. Terns of San Diego County. 29 March 2007. Presented at the San Diego Audubon Society meeting, Tecolote Canyon Nature Center.

"Terns" Chesapeake Bay Program. Bay Field Guide. 25 August 2008 <http://www.chesapeakebay.net/bfg_terns.aspx?menuitem=19352>.

Toropova, C. "Rasa Island." Blue Voice. 19 March 2007 <http://www.bluevoice.org/sections/ocean/rasa.shtml>.

Unitt, P. "Elegant Tern." *San Diego County Bird Atlas*. San Diego: Ibis Publishing, 2004.

Velarde, E., et al. 2004. Seabird Ecology, El Niño Anomalies, and Prediction of Sardine Fisheries in the Gulf of California. *Ecological Applications,* (14)2:607–15.

West, L., and P. Unitt. "Elegant Terns." Ocean Oasis: Field Guide. San Diego Natural History Museum. 27 February 2007. <http://www.oceanoasis.org/fieldguide/ster-ele.html>.

Gull-Billed Tern

"Gull Billed Tern: *Sterna Nilotica*." U.S. Geological Survey. 27 February 2007. <http://www.mbr-pwrc.usgs.gov/id/framlst/i0630id.html>.

Molina, K.C., and M.R. Erwin. 2006. The Distribution and Conservation Status of the Gull-billed Tern (*Gelochelidon nilotica*) in North America. *Waterbirds,* 29(3):561–83.

Molina, K.C., and D.A. Marschalek. 2003. "Foraging Behavior and Diet of Breeding Western Gull-Billed Terns (*Sterna nilotica vanrossemi*) in San Diego Bay." Species Conservation and Recovery Program Report 2003-01. California Department of Fish and Game, Habitat Conservation, Sacramento, CA.

Robbins, C.S., B. Bruun, and H.S. Zim. *Birds of North America*. New York: Western Publishing Company, Inc., 1966.

Unitt, P. "Gull Billed Tern." *San Diego County Bird Atlas*. San Diego: Ibis Publishing, 2004.

U.S. Fish and Wildlife Service. 1918. *Migratory Bird Treat Act of 1918*. Digest of Federal Resource Laws of Interest to the U.S. Fish and Wildlife Service.

Light-Footed Clapper Rail

Audubon, J.J. *Birds of America*. USA: Publisher unknown, 1842.

"About Our Cause." Clapper Rail Study Team. 24 May 2007. <http://clapperrail.com/about.htm>.

Gailband, C. "Restoring a Local Endangered Species." 19 May 2007. Lecture, Imperial Beach, CA.

Mannes, T. "Rare birds get helping hand." *San Diego Union-Tribune*. 8 September 2007.

Unitt, P. "Clapper Rail." *San Diego County Bird Atlas*. San Diego: Ibis Publishing, 2004.

Vanner, M. *The Encyclopedia of North American Birds*. Bath, England: Parragon, 2002.

Zedler, J.B. *The Ecology of Tijuana Estuary*. San Diego: San Diego State University, 1992.

Zembal, R., S. Hoffmann, and J. Konecny. *Status and Distribution of the Light-footed Clapper Rail in California, 2007*. Report to the California Department of Fish and Game, South Coast Region and the U.S. Fish and Wildlife Service, 2007. 14 pp.

Black Oystercatcher

Andres, B.A. 1999. Effects of Persistent Shoreline Oil on Breeding Success and Chick Growth in Black Oystercatchers. *The Auk*, 116(3):640-50.

"Black Oystercatcher *Haematopus bachmani*." *The Pacific Wildlife Foundation*. 23 February 2007. <http://www.pwlf.org/blackoystercatcher.htm>.

Tessler, D. F., et al. "Black Oystercatcher (*Haematopus bachmani*) Conservation Action Plan. 1 April 2007." Alaska Department of Fish and Game. 10 June 2008. <http://www.fws.gov/oregonfwo/Species/Data/BlackOystercatcher/ Documents/Black_oystercatcher_conservation_action_plan_FINAL_April07. pdf>.

Western Snowy Plover

Hornaday, K., I. Pisani, and B. Warne. 2007. "Recovery Plan for the Pacific Coast Population of the Western Snowy Plover (*Charadrius alexandrinus nivosus*)." (2 volumes) Sacramento, CA: U.S. Fish and Wildlife Service. xiv + 751 pp.

"The Snowy Plover Page." Friends of the Dunes. 20 March 2007. <http://www. friendsofthedunes.org/nature/western-snowy-plover.shtml>.

"Snowy Plover." National Audubon Society. 15 March 2007. <http://web1.audubon. org/science/species/watchlist/index.php>.

"Tijuana River National Estuarine Research Reserve." 31 May 2007. <http://trnerr.org/ endang.html>.

Unitt, P. "Western Snowy Plover." *San Diego County Bird Atlas*. San Diego: Ibis Publishing, 2004.

U.S. Fish and Wildlife Service. 2007. Recovery Plan for the Pacific Coast Population of the Western Snowy Plover (*Charadrius alexandrinus nivosus*). In 2 volumes. Sacramento, CA. xiv + 751 pages.

Black Skimmer

"All About Birds: Black Skimmer." Cornell Lab of Ornithology. 27 February 2007. <http://www.birds.cornell.edu/AllAboutBirds/BirdGuide/Black_Skimmer_dtl. html>.

"Black Skimmer *Rynchops niger*." The University of Georgia Museum of Natural History. 27 February 2007. <http://dromus.nhm.uga.edu/~GMNH/gawildlife/index. php?page=speciespages/species_page&key=rniger>.

Collins, C. "Black Skimmer (*Rhynchops niger*)." U.S. Fish and Wildlife Service. 27 February 2007. <http://www.fws.gov/bolsachica/BlackSkimmerprofile.htm>.

Safina, C., and J. Burger. 1983. Effects of Human Disturbance on Reproductive Success in the Black Skimmer. *The Condor*, 85(2):164–71.

Unitt, P. "Black Skimmer." *San Diego County Bird Atlas*. San Diego: Ibis Publishing, 2004.

Long-Billed Curlew

"Curlew." The Royal Society for the Protection of Birds. 3 March 2007. <http://www.rspb.org.uk/wildlife/birdguide/name/c/curlew/index.asp>.

Dugger, B.D., and K.M. Dugger. "Long-billed Curlew" The Birds of North America Online. 22 August 2008. <http://bna.birds.cornell.edu/bna/species/628/articles/introduction>.

"The Long-billed Curlew." National Audubon Society. 11 May 2007. <http://www.audubon.org/bird/BoA/F36_G10a.html>.

Unitt, P. "Long-billed Curlew." *San Diego County Bird Atlas*. San Diego: Ibis Publishing, 2004.

Osprey

Kirshbaum, K., and P. Watkins. 2000. *"Pandion haliaetus"* (Online) Animal Diversity Web. Accessed 26 February 2007. <http://animaldiversity.ummz.umich.edu/site/accounts/information/Pandion_haliaetus.html>.

Kucher, K. "Entangled bird of prey freed from light pole." 13 June 2006. *San Diego Union-Tribune*. Accessed 3 June 2007. <http://www.signonsandiego.com/news/metro/20060613-1337-bn13bird.html>.

Magee, M. "$5 million tagged for environmental fund." 10 September 2006. *San Diego Union-Tribune*. Accessed 21 February 2007. <http://weblog.signonsandiego.com/uniontrib/20060910/news_1m10port.html>.

—. "Nesting platforms to be built for ospreys." 5 April 2007. *San Diego Union-Tribune*. Accessed 27 April 2007. <http://www.signonsandiego.com/news/metro/20070405-9999-1m5bird.html>.

"Osprey: General Info." Chesapeake Bay Program. 26 February 2007. <http://www.chesapeakebay.net/bfg_osprey.aspx?menuitem-14381>.

"Osprey." National Audubon Society. 3 April 2007. <http://web1.audubon.org/waterbirds/species.php?speciesCode=osprey>.

"Osprey *Pandion haliaetus* Identification Tips." U.S. Geological Survey. 26 February 2007. <http://www.mbr-pwrc.usgs.gov/id/framlst/i3640id.html>.

Unitt, P. "*Pandion haliaetus* Osprey. 2000" Ocean Oasis Field Guide. San Diego Natural History Museum. 22 February 2007. <http://www.oceanoasis.org/fieldguide/pand-hal.html>.

Peregrine Falcon

Lincer, J.L. 1975. DDE-Induced Eggshell-Thinning in the American Kestrel: A Comparison of the Field Situation and Laboratory Results, *J. Appl. Ecol.* 12(3):781–93.

Pavelka, M.A. 1990. Peregrine Falcons Nesting in San Diego, California. *Western Birds*, 21:181–83.

"Peregrine Falcon." Cornell Lab of Ornithology. 24 February 2007. <http://www.birds.cornell.edu/AllAboutBirds/BirdGuide/Peregrine_Falcon.html>.

"Peregrine Falcon[2]." The Peregrine Fund. 24 February 2007. <http://www.peregrinefund.org/explore_raptors/falcons/peregrin.html>.

"Peregrine Falcon[3]." U.S. Geological Survey. 24 February 2007. <http://www.mbr-pwrc.usgs.gov/id/framlst/i3560id.html>.

"Peregrine Falcon[4]." U.S. Fish and Wildlife Service. 24 February 2007 <http://ecos.fws.gov/speciesProfile/SpeciesReport.do?spcode=B050>.

"Peregrine Falcon (*Falco peregrinus*)." 2006. U.S. Fish and Wildlife Service. <library.fws.gov/ES/peregrine06.pdf>.

"Peregrine Falcons and DDT. 1998–2007." Santa Cruz Predatory Bird Research Group.

6 February 2007. <http://www2.ucsc.edu/scpbrg/ddt.htm>.

"Raptor Identification" U.S. Department of the Interior. Bureau of Land Management. 24 February 2007 <http://www.blm.gov/id/st/en/fo/four_rivers/01/links/raptor_identification.html>.

Sooter, W. *Personal Interview.* 2007.

Unitt, P. "Peregrine Falcon." *San Diego County Bird Atlas.* San Diego: Natural History Museum, Ibis Publishing, 2004.

Walton, B.J. 2006. "The History of SCPBRG." 29 August 2008. <http://www2.ucsc.edu/scpbrg/founding.htm>.

White, C.M., et al. "Peregrine Falcon." The Birds of North America Online. 23 February 2007. <http://bna.birds.cornell.edu/bna/species/660/articles/introduction>.

"Zoological Society's Highest Honor Goes to Three Conservationists." 2 October 2006. Conservation and Research for Endangered Species (CRES). Accessed 26 June 2009. <http://fionkiro.blogspot.com/2006/09/conservation-news.html>.

Northern Harrier

"Bay-Delta Region River Report: Northern Harrier." California Department of Fish and Game. 11 June 2008. <http://www.delta.dfg.ca.gov/reports/stanriver/sr4313.asp>.

"*Circus Cyaneus* Linneaus: Northern Harrier." July 2001. Michigan Natural Features Inventory, Michigan State University Extension. 10 June 2008. <http://web4.msue.msu.edu/mnfi/abstracts/zoology/Circus_cyaneus.pdf>.

Ehrlich, P.R., D.S. Dobkin, and D. Wheye. 1988. "DDT and Birds." Stanford University. 23 May 2007. <http://www.stanford.edu/group/stanfordbirds/text/essays/DDT.html>.

—. "The Blue List." 1988. Stanford University. Accessed 23 May 2007. <http://www.stanford.edu/group/stanfordbirds/text/essays/The_Blue_List.html>.

"Northern Harrier" Bird Web, Seattle Audubon Society. 10 June 2008. <http://birdweb.org/birdweb/bird_details.aspx?id=99>.

"The Northern Harrier: *Circus cyaneus.*" Great Salt Lake Playa Foodweb Project. 26 February 2007. <http://people.westminstercollege.edu/faculty/tharrison/gslplaya99/harrier.htm>.

"Northern Harrier Fact Sheet." New York State, Department of Environmental Conservation. 26 February 2007. <http://www.dec.ny.gov/animals/7090.html>.

Remsen, J.V., Jr. 1978. "Bird Species of Special Concern in California." Report 78-1. 54 pp. California Department of Fish and Game, Wildlife Management Administrative, Sacramento, CA.

Unitt, P. "Northern Harrier." *San Diego County Bird Atlas.* San Diego: Ibis Publishing, 2004.

Burrowing Owl

Abbott, C.G. 1930. Urban Burrowing Owls. *The Auk,* 47:564–65.

"Burrowing Owl: An Endangered Species." 2001. Saskatchewan Schools and School Divisions. 4 June 2007. <http://www.saskschools.ca/~gregory/animals/burowl1.html>.

Gailband, C. "Restoring a Local Endangered Species." 19 May 2007. Lecture, Imperial Beach, CA.

Lincer, J. 2005 Burrowing Owl Program Update. Wildlife Research Institute. 25 March 2008. <www.wildlife-research.org/wildlife_news_2005.pdf>

"Owl Biology." Saskatchewan Burrowing Owl Interpretive Center. 25 February 2007. <http://www.sboic.ca/biology.php>.

Restani, M. 2001. Nest Site Selection and Productivity of Burrowing Owls Breeding on the Little Missouri National Grassland. USDA Forest Service, Washington, D.C.

Sheffield, S.R. 1997. Current Status, Distribution, and Conservation of the Burrowing

Owl (*Speotyto cunicularia*) in Midwestern and Western North America. In: *Biology and Conservation of Owls of the Northern Hemisphere*. J.R. Duncan, D.H. Johnson, and T.H. Nicholls, eds. U.S. Forest Service Gen. Tech. Rep. NC-190. pp. 399–407.

Unitt, P. "Burrowing Owl." *San Diego County Bird Atlas*. San Diego: Ibis Publishing, 2004.

Wong, K. 2004. "Spread the Dung, It's Time for Dinner." *Science Now*. <http://sciencenow.sciencemag.org/cgi/content/full/2004/901/3?etoc>.

Gray Whale

Angliss, R.P., and R.B. Outlaw. 2005. Gray Whale (*Eschrichtius robustus*): Eastern North Pacific Stock. Alaska Marine Mammal Stock Assessments, 2005. <http://www.nmfs.noaa.gov/pr/pdfs/sars/ak2005whgr-en.pdf>.

Dana, R.H. *Two Years Before the Mast*. New York: Penguin/Viking, 1840.

Gordon, D., and A. Baldridge. *The Gray Whale*. Monterey, CA: Monterey Bay Aquarium, 1991.

"Gray Whale." National Geographic. 24 February 2007. <http://www3.nationalgeographic.com/animals/mammals/gray-whale.html>.

"Gray Whale Migration Route." Journey North. 11 June 2008 <http://www.learner.org/jnorth/tm/gwhale/MigrationRoute_Map.html>.

"Gray Whale Tutorial." Orca Network. 11 June 2008. <http://orcanetwork.org/nathist/graywhales.html>.

Gregr, E.J., et al. 2000. Migration and Population Structure of Northeastern Pacific Whales Off Coastal British Columbia: An Analysis of Commercial Whaling Records From 1908–1967. *Marine Mammal Science*, 16(4):699–727.

Heyning, J.E., and C.A. Heyning. 2001. Rescue of an Orphaned Gray Whale Calf. *Aquatic Mammals*, 27(3):212-14.

May, R.V. "Dog-Holes, Bomb-Lances and Devil-Fish: Boom Times for the San Diego Whaling Industry." The Journal of San Diego History. Volume 32, Spring 1986: Number 2. 10 June 2008. <http://www.sandiegohistory.org/journal/86spring/dogholes.htm>.

May, R.V. *Personal Interview*. 2007.

McClure, R. "Judge puts limits on Navy sonar use." 4 January 2008. *Seattle PI*. 11 June 2008. <http://seattlepi.nwsource.com/local/346011_orcasonar04.html>.

Perryman, W.L., et al. 2002. Gray Whale Calf Production 1994–2000: Are Observed Fluctuations Related to Changes in Seasonal Ice Cover? *Marine Mammal Science*, 18(1):121–44.

Rice, D.W., and A.A. Wolman. 1971. The Life History and Ecology of the Gray Whale (*Eschrichtius robustus*). *Am. Soc. Mammal. Spec. Publ.*, 3:1–142.

"Save the Whales." 12 December 2007. <http://savethewhales.org/about.html>.

Sidenstecker, M. *Personal Interview*. 2007.

"US Navy in sonar ban over whales." 4 July 2006. British Broadcasting Corporation. 12 December 2007. <http://news.bbc.co.uk/2/hi/americas/5143698.stm>.

Watson, L. *Sea Guide to Whales of the World*. New York: Dutton, 1981.

California Sea Otter

"Behavior of Sea Otter." 2006. Friends of the Sea Otter. 26 February 2007. <http://www.seaotters.org/geteducated.html>.

Conrad, P.A., et al. 2005. Transmission of Toxoplasma: Clues from the Study of Sea Otters as Sentinels of *Toxoplasma gondii* Flow into the Marine Environment. *Int. J. Parisitology*, 35:1125–68.

Estes, J.A. 2004[1]. Research and Conservation of Sea Otters in California. Presentation. Alaska Sea Otter Research Workshop. Alaska Sea Grant College Program, Fairbanks. pp. 10–11.

——. 2004[2]. The Ecology and Population Biology of Sea Otters. An Overview. Presentation. Alaska Sea Otter Research Workshop. Alaska Sea Grant College Program, Fairbanks. pp. 7–8.

Estes, J.A. 2002. "What's Wrong with the California Sea Otter?" U.S. Geologic Survey. Sound Wave. 26 February 2007. <http://soundwaves.usgs.gov/2002/02/research.html>.

"Sea Otter." 14 December 1998. James Ford Bell Library. 26 February 2007. <http://www.bell.lib.umn.edu/Products/SeaOtter.html>.

"Sea Otter MMC." March 2002. The Marine Mammal Center. 26 February 2007. <http://www.tmmc.org/pdfs/library/Sea_Otter.pdf>.

Skinner, J.E. 1962. The Mammalian Resources. In: *An Historical Review of the Fish and Wildlife Resources of the San Francisco Bay Area.* Skinner, J.E., ed. California Department of Fish and Game Water Projects Branch Report No. 1.

Story of the Sweetwater Rainbow Trout

Rodgers, T. "Enthusiastic efforts, disheartening results." *San Diego Union-Tribune,* March 11, 2007. <http://cfx.signonsandiego.com/news/metro/20070311-9999-lz1n11trout.html>.

Ship Traffic

Chase, J.S. 1999. Recreation. *California Coast and Ocean,* June 1999. pp. 10–14.

Dana, R.H. *Two Years Before the Mast.* New York: Penguin/Viking, 1840.

Merk, D. *Personal Interview.* 15 March 2007.

"San Diego History Timeline." San Diego Historical Society. 8 March 2007. <http://www.sandiegohistory.org/timeline/timeline1.htm#1760>.

"Stop Aquatic Invaders on Our Coast!" 2006. (Bilingual Poster) Sea Grant Extension Program. <http://www.csgc.ucsd.edu/NEWSROOM/NEWSRELEASES/SAI_.html>.

Invasive Species

Carlton, J.T. *Introduced Species in U.S. Coastal Waters.* Arlington, VA: Pew Oceans Commission, 2001.

Carlton, J.T., and V.A. Zullo. 1969. Early Records of the Barnacle *Balanus improvisus* Darwin on the Pacific Coast of North America. *Occ. Pap. Calif. Acad. Sci,* 75:6.

Culver, C.S., A.M. Kuris, and B. Beede. *Identification and Management of the Exotic Sabellid Pest in California Cultured Abalone.* La Jolla, CA: California Sea Grant College System, 1997.

Diamond, J. *Collapse.* New York: Penguin, 2005.

Gonzalez, J. *Personal Interview.* March 2007.

Hyman, L.H. 1955. The Polyclad Flatworms of the Pacific Coast of North America: Additions and Corrections. *Amer. Mus. Novitates,* 1704:11.

Miller, R.L. 1969. *Ascophyllum nodosum*: A Source of Exotic Invertebrates Introduced into West Coast Near-Shore Marine Waters. *Veliger,* 12:230–31.

Van Heertum. R. 2002. Introduced Species Summary Project. European Green Crab *Carcinus maenas*). 29 August 2008. <http://www.columbia.edu/itc/cerc/danoff-burg/invasion_bio/inv_spp_summ/Carcinus_maenas.htm>.

Wasson K., et al. 2001. Biological Invasions of Estuaries Without International
Shipping: The Importance of Intraregional Transport. *Biological Conservation*,
102(2):143–53.

Woodfield, R. 2000. Noxious Seaweed Found in Southern California Coastal Waters.
Agua Hedionda. *Caulerpa Taxifolia* Reports. 29 August 2008. <http://swr.nmfs.
noaa.gov/hcd/caulerpa.htm>.

Dredging

Cooper, H.R. *Practical Dredging*. Glasgow: Brown, Son & Ferguson, Ltd., 1958.

Herbich, J.B. *Coastal and Deep Ocean Dredging*. Houston: Gulf Publishing Company, 1975.

—. *Handbook of Dredging Engineering*. New York: McGraw-Hill, Inc., 1992.

Marcus L. *The Coastal Wetlands of San Diego County*. Sacramento, CA: California State
Coastal Conservancy, 1989.

Meegoda, J.N., et al. *Dredging and Management of Dredged Materials*. Reston, VA: The
American Society of Civil Engineers, 1997.

Perdue, M. *Personal Interview*. 2007.

Unitt, P. *San Diego County Bird Atlas*. San Diego: Ibis Publishing, 2004.

Land Use

Carson, R. *Silent Spring*. Boston: Mariner Books, 2002.

Gustaitis, R. 2003. "Sterilizing Creek Water." *California Coast and Ocean*. <http://ceres.
ca.gov/coastalconservancy/coast&ocean/winter2002-03/pages/two_sidebarA.
htm/>.

High Tech High. *San Diego Bay: A Story of Exploitation and Restoration*. La Jolla, CA:
California Sea Grant College Program, 2007.

"The Jazz Age: The American 1920s." Digital History. 14 May 2007. <http://www.
digitalhistory.uh.edu/database/article_display_printable.cfm?HHID=454>.

"Kumeyaay History." 15 May 2007. <http://www.viejasbandofkumeyaay.org/html/
tribal_history/kumeyaay_history.html>.

"Land Use—Regional Growth." SANDAG. 14 May 2007 <http://www.sandag.cog.
ca.us/index.asp?classid=12&fuseaction=home.classhome>.

Lee, M. "S.D. Bay cleanup mandate stagnates." *San Diego Union-Tribune*. 30 April 2007.
<http://www.signonsandiego.com/news/metro/20070430-9999-1n30bay.html>.

Mannes, T. "Port Commission rejects smaller power plant on bayfront." *San Diego
Union-Tribune*. 14 March 2007. <http://www.signonsandiego.com/news/
metro/20070314-9999-1m14plant.html>.

May, R.V. "Dog-Holes, Bomb-Lances and Devil-Fish: Boom Times for the San Diego
Whaling Industry." The Journal of San Diego History. Volume 32, Spring 1986:
Number 2. 10 June 2008. <http://www.sandiegohistory.org/journal/86spring/
dogholes.htm>.

McKeever, M. *A Short History of San Diego*. San Diego: Lexikos, 1985.

Mills, J.R. "San Diego—Where California Began. Part 3—Mexican Interlude."
1985. San Diego Historical Society. Accessed 16 May 2006. <http://www.
sandiegohistory.org/books/wcb/wcb3.htm>.

"Naval Training Center." San Diego Navy Historical Association. 17 May 2007. <http://
www.quarterdeck.org/AreaBases/NTC%20History_files/ntc_history.htm>.

Perdue, M. *Personal Interview*. 2007.

Port of San Diego[1]. "Jurisdictional Urban Runoff Management Program Document,
2005." 14 May 2007. <http://www.portofsandiego.org/environment/
stormwater/304-jurisdictional-urban-runoff-management-program-jurmp-
document.html>.

Port of San Diego[2]. "Natural Resources Management Plan, 1999." 14 May 2007. <http://www.portofsandiego.org/environment/natural-resources/312-natural-resources-management-plan.html>.

Port of San Diego[3]. "Natural Resources." 14 May 2007. <http://www.portofsandiego.org/environment/natural-resources.html>.

"Port SUSMP." Port of San Diego. 14 May 2007. <http://www.portofsandiego.org/sandiego_environment/susmp.asp>.

Reznik, B. *Personal Interview.* 29 March 2007.

"Timeline of San Diego History." San Diego Historical Society. 13 May 2007. <http://www.sandiegohistory.org/timeline/timeline.htm>.

Shragge, A.J. "I Like the Cut of Your Jib: Cultures of Accommodation Between the U.S. Navy and Citizens of San Diego, California, 1900–1951." The Journal of San Diego History. Volume 48, Summer 2002: Number 3. 10 June 2008. <http://www.sandiegohistory.org/journal/2002-3/navy.htm>.

"UC IPM Online." University of California Agricultural and Natural Resources. 13 May 2007. <http://www.ipm.ucdavis.edu/GENERAL/urbanpesticideuse.html>.

"UC Toxics News" Fall/Winter 2000. University of California Toxic Substances Research & Teaching Program. 14 May 2007. <http://www.tsrtp.ucdavis.edu/public/our_program/newsletters/Fall_Winter_00/homepage.php>

Climate Change

Bell, D. "Bay once was a whale of a nursery. "*San Diego Union-Tribune.* 25 March 2006. <http://www.signonsandiego.com/uniontrib/20060325/news_1m25bell.html>.

"Climate Change Impacts on the United States: The Potential Consequences of Climate Variability and Change." U.S. Climate Change. Science Program. 21 April 2007. <http://www.usgcrp.gov/usgcrp/Library/nationalassessment/overview.htm>.

Durkin, M. *The Great Global Warming Swindle.* Channel 4 Television Corporation, U.K. 8 March 2007. <http://www.channel4.com/science/microsites/G/great_global_warming_swindle/>.

Haymet, A.D.J. *Personal Interview.* 2007.

"Hot Politics of Global Warming." Public Broadcasting Service Frontline. 24 April 2007. <http://www.pbs.org/wgbh/pages/frontline/hotpolitics/>.

Karling, H. *Global Climate Change.* Huntington, NY: Nova Science Publishers, 2001.

Krier, R. "Global warming's link to wacky weather cloudy." 4 March 2007. *San Diego Union-Tribune.* Accessed 12 December 2007. <http://www.signonsandiego.com/news/science/20070304-9999-1m4warm.html>.

Landler, M. "Bush's Climate Plan Alters Showdown with Europe." *New York Times.* 1 June 2007. Accessed 24 April 2007. <http://www.nytimes.com/2007/06/01/world/europe/01cnd-germany.html?_r=1&oref=slogin>.

Oppenheimer, M. *Personal Interview.* San Diego Natural History Museum. 15 May 2007.

Roleeff, T. *Opposing Viewpoints.* San Diego: Greenhaven Press, 1997.

Rugh, S. *Personal Interview.* San Diego Natural History Museum. 3 May 2007.

Story of the Baiji River Dolphin

Massicot, P. "Animal Info—Baiji." 4 June 2006. <http://www.animalinfo.org/species/cetacean/lipovexi.htm>.

Stewart, B.S. 2007. Presentation. Hubbs SeaWorld Research Institute.

U.S. Geological Survey. *Minerals Yearbook—Cement.* Washington, D.C.: U.S. Department of the Interior, 2006.

"Yangtze Freshwater Dolphin Expedition 2006." baiji.org Foundation. <http://www.baiji.org/expeditions/1.html>.

Endangered Species Act

"Endangered and Threatened Bird Species at the Tijuana Estuary." Tijuana River National Estuarine Research Reserve. 28 May 2007. <http://trnerr.org/endang. html>.

"The Endangered Species Act." Natural Resources Defense Council. 21 May 2007 <http://www.nrdc.org/wildlife/habitat/esa/aboutesa.asp>.

"The Endangered Species Act: A Backgrounder." Environmental Defense Network. 21 May 2007 <http://www.edf.org/article.cfm?contentID=3686>.

Ezcurra, E. *Personal Interview*. San Diego Natural History Museum. 3 April 2007.

"Habitat Conservation Plans. Working Together for Endangered Species." 12 December 2007. U.S. Fish and Wildlife Service. <http://www.fws.gov/endangered/pubs/ HCPBrochure/HCPsWorkingTogether5-2005web%20.pdf>.

Kostel, K. "A Top Predator Roars Back." Summer 2004. Natural Resources Defense Council. 10 June 2008. <http://www.nrdc.org/onearth/04sum/briefings.asp>.

O'Toole, R. "The History of the Endangered Species Act." Winter 1996. Electronic Drummer. 10 June 2008. <http://www.ti.org/ESAHistory.html>.

"The Peregrine Fund." World Center for Birds of Prey. 25 April 2007. <http://www. peregrinefund.org/default.asp>.

Sherry, C.J. *Endangered Species: A Reference Handbook*. Santa Barbara, CA: ABC-CLIO, Inc., 1998.

Shogren, J. 2005. *Species at Risk: Using Economic Incentives to Shelter Endangered Species on Private Lands*. Austin, TX: University of Texas Press, 2005.

"Species Profile: Light-footed clapper rail." 21 May 2007. U.S. Fish and Wildlife Service. <http://ecos.fws.gov/speciesProfile/SpeciesReport.do?spcode=B04B>.

"Species Profile: Western snowy plover." 21 May 2007. U.S. Fish and Wildlife Service. <http://ecos.fws.gov/speciesProfile/SpeciesReport.do?spcode=B07C>.

"The Status of Rare, Threatened, and Endangered Plants and Animals of California 2000–2004." California Department of Fish and Game. 2005. 10 June 2008. <http://www.dfg.ca.gov/wildlife/nongame/t_e_spp/new_te_rpt.html>.

Suckling, K., R. Slack, and B. Nowicki. "Extinction and the Endangered Species Act." Center for Biological Diversity. 1 May 2004. <http://www.biologicaldiversity.org/ swcbd/Programs/policy/esa/EESA.pdf>.

Stewards of the Bay

"About Pro Peninsula." 2006. Pro Peninsula. 28 February 2007. <http://www. propeninsula.org/content/1/1/1.html>.

"About the Nature Center." 2002. Chula Vista Nature Center. <http://www. chulavistanaturecenter.org/About/default.asp>.

Browning, B.M., J.W. Speth and W. Gayman. *The Natural Resources of San Diego, CA. Bay: Their Status and Future*. Sacramento: California Department of Fish and Game, 1973.

"California Coastal Commission: Why it exists and what it does." 2005. California Coastal Commission. 10 June 2008. <http://www.coastal.ca.gov/publiced/ Comm_Brochure.pdf>.

"Coastkeeper's Litigation Campaign." 2006. *Coastkeeper*. 10 June 2008. <http://www. sdcoastkeeper.org/content/lawPolicy/campaigns/stormWater/cmp-str_shipyards. htm>.

"CRES About Us" 2007. Conservation and Research for Endangered Species. 6 March 2007. <http://www.sandiegozoo.org/conservation/about>.

Crooks, J.A. Invasions and Effects of Exotic Marine Species: A Perspective from Southern California. American Fisheries Society Meeting, Monterey, CA, 1997.

"Famosa Slough." 2006. Friends of Famosa Slough. 6 March 2007. <http://www.famosaslough.org/>.

Helly, J.N., et al. 2001. Collaborative Management of Natural Resources in San Diego Bay. *Coastal Management*, 29(2) 117–32.

High Tech High. *Perspectives of the San Diego Bay: A Field Guide*. Providence, RI: Next Generation Press, 2005.

Moore, B. *Personal Interview*. 2006.

Reznik, B. *Personal Interview*. 29 March 2007.

Rodgers, T. "S.D. water-quality board goes from 'worst to best'." *San Diego Union-Tribune*. 17 February 2006. <http://www.signonsandiego.com/news/metro/20060217-9999-1n17board.html>.

"San Diego Bay National Wildlife Refuge Complex." U.S. Fish & Wildlife Service. 12 March 2008. <http://www.fws.gov/sandiegorefuges/>.

"State of the Bay: January 2007." Unified Port of San Diego. 17 January 2007. <http://www.portofsandiego.org/sandiego_environment/documents/State_of_the_Bay.pdf>.

"SDAS Conservation Projects." San Diego Audubon Society. 28 February 2007. <http://www.sandiegoaudubon.org/conservation.htm>.

"Southern California Bight 2008 Regional Monitoring Program Documents." Southern California Coastal Water Research Project. 12 March 2008. <http://www.sccwrp.org/view.php?id=524>

"Wildcoast Annual Report." 2006. Accessed 29 August 2008. <http://www.wildcoast.net/site/index.php?option=com_content&task=view&id=351&Itemid=70 p. 270>.

Eelgrass Restoration

Anderson, T. and K. Hovel. "Evaluating Eelgrass Restoration: Effects of Habitat Structure on Fish Recruitment and Epifaunal Diversity in San Diego Bay." Unified Port of San Diego. 3 February 2005. <www.portofsandiego.org/docman/doc_download/313-eel-grass-restoration-study-sdsu-report-.html>.

"City of San Diego Water History." City of San Diego Water Department. 26 April 2007 <http://www.sandiego.gov/water/gen-info/history.shtml>.

"Coronado Bridge Seismic Retrofit Project." January 2002 Fact Sheet. State of California—Business, Transportation and Housing Agency, Department of Transportation, District 11, San Diego, CA. 29 August 2008. <http://www.dot.ca.gov/dist11/facts/coronado.htm>.

"Eelgrass Update." City of Newport Beach, California. 26 April 2007 <http://www.city.newport-beach.ca.us/HBR/Eelgrass%20Update%208-03.pdf>.

Hovel, K. *Personal Interview*. 2007.

"Management of Environmental Mitigation." Unified Port of San Diego. 6 June 2006. <www.portofsandiego.org/about-us/bpc-policies/doc_download/1417-bpc-policy-no-735-environmental-mitigation-property.html>.

"Natural Resources." Unified Port of San Diego. 26 April 2007. <www.portofsandiego.org/docman/doc_download/1717-sdsu-maintaining-healthy-eelgrass-beds-t-progress-july-september-2008.html>.

"San Diego Bay Integrated Natural Resources Management Plan." U.S. Department of the Navy, Southwest Division (USDoN, SWDIV), San Diego, CA. September 2000.

"Southern California Eelgrass Mitigation Policy." National Oceanic and Atmospheric Administration. 31 July 1991. <http://swr.nmfs.noaa.gov/hcd/policies/EELPOLrev11_final.pdf>

"State of the Bay: January 2007." Unified Port of San Diego. 17 January 2007. <http://www.portofsandiego.org/sandiego_environment/documents/State_of_the_Bay.pdf>.

Williams, S. 2001. Reduced genetic diversity in eelgrass transplantations affects both individual and population fitness. *Ecological Applications,* 11:1472–88.

Sustainable Fisheries

Blackburn, T.C., and K. Anderson. *Before the Wilderness.* Menlo Park, CA: Ballena Press, 1993.

Brown, L.R. 2001. "Eco-Economy: Building an Economy for the Earth." Earth Policy Institute. 17 March 2007 <http://www.earth-policy.org/Books/Eco/EEch3_ss2.htm>.

Diamond, J. *Guns, Germs, and Steel: The Fates of Human Societies.* New York: W.W. Norton & Company, 1997.

Eilperin, J. "House Approves Overhaul of Rules for Fisheries." *Washington Post.* 10 December 2006. 17 March 2007 <http://www.washingtonpost.com/wp-dyn/content/article/2006/12/09/AR2006120900818.html>.

High Tech High. *San Diego Bay: A Story of Exploitation and Restoration.* pp. 148–50. La Jolla: California Sea Grant College Program, 2007.

"Introduction to Eco-Labeling." Global Eco-Labeling Network. July 2004. Accessed 17 March 2007. <http://www.gen.gr.jp/pdf/pub_pdf01.pdf>.

Neushul, P. 1989. Seaweed for War: California's World War I Kelp Industry. *Technology and Culture,* 30(3): 561–83.

"NOAA Aquaculture Program." 7 March 2007. <http://aquaculture.noaa.gov/>.

Rodgers, T. "Foundation releases 2,000 white sea bass into Mission Bay to keep up replenishing." 14 November 2006. *San Diego Union-Tribune.* Accessed 17 March 2007. <http://www.signonsandiego.com/sports/outdoors/20061114-9999-1m14bass.html>.

Speer, L., et al. *Hook, Line, and Sinking: The Crisis in Marine Fisheries.* New York: National Resources Defense Council, 1997.

Marine Protected Areas

Arnett, W.A. 2006 "Earth." The Nine Planet Solar System Tour. <http://www.nineplanets.org/earth.html>.

Becker, B.J. "Status and Trends of Ecological Health and Human Use of the Cabrillo National Monument Rocky Intertidal Zone (1990–2005)." Natural Resource Technical Report NPS/PWR/CABR/NRTR 2006/03. Seattle: National Park Service, 2006. 195 pp.

"China." CIA World Factbook. 15 May 2008. Accessed 10 June 2008. <https://www.cia.gov/library/publications/the-world-factbook/geos/ch.html>.

Compton, A. *Personal Interview.* 2007.

Gage, B. *Personal Interview.* December 2006.

Häder, D.-P. *The Effects of Ozone Depletion on Aquatic Ecosystems.* Environmental Intelligence Unit. Austin, TX: R.G. Landes Company, 1997.

McArdle, D.A. *California Marine Protected Areas Past and Present.* La Jolla, CA: California Sea Grant College Program, 2002.

"MPA Connections." National Marine Protected Areas Center. August 2005. <http://www.mpa.gov/pdf/helpful-resources/connections/connections-aug-sept05.pdf>.

——. "Level of Protection." 2003. <www.mpa.gov/.../factsheets/mma_analysis_oct06.pdf>.

"NOAA, AOML State of the Ocean." National Oceanic and Atmospheric Administration. Atlantic Oceanographic and Meteorological Laboratory. 17 January 2007 <http://www.aoml.noaa.gov/phod/soto/index.php>.

"NOAA." Office of Communications. 12 December 2007 <http://www.noaa.gov/about-noaa.html>.

Norse, E., and L. Crowder. *Marine Conservation Biology the Science of Maintaining the Sea's Biodiversity*. Washington, D.C.: Island Press, 2005.

Smith, F. *Environmental Sustainability*. USA: CRC Press, 1997.

Sobel, J., and C. Dahlgren. *Marine Reserves: A Guide to Science, Design, and Use*. Washington, D.C.: Island Press, 1998.

"South China Sea." Sanctuaries Web Team. 24 April 2007. <http://sanctuaries.noaa. gov/management/international/mpa_schina3.html>.

"United States." 15 May 2008. Accessed 10 June 2008. <https://www.cia.gov/library/ publications/the-world-factbook/geos/us.html>.

Carl Hubbs

"Biography on the History of Scripps Institution of Oceanography." (Rev. April 7, 2005). 12 March 2008. <repositories.cdlib.org/cgi/viewcontent. cgi?article=1007&context=sio/arch>

Campbell, S. "Carl Leavitt Hubbs (1894–1979)." *ZOONOOZ*, 1979. pp. 14–15.

Cox, A. 1981 "Laura Hubbs At His Side—Not In His Shadow" *Witnesses*, May 1981. pp.11–15.

"Historical Perspectives … A Name to Build On." Hubbs-SeaWorld Research Institute. 1963. *Currents* Number 53.

Horn, M.H. 1976. In Honor of Carl L. Hubbs. *Bulletin of Southern California Academy of Sciences* 75(2) 57–9.

"Laura Clark Hubbs, Scripps Institution associate, dies at 95." *San Diego Union-Tribune*. 1988.

Rosenblatt, R.H. "Carl Leavitt Hubbs, 1894–1979." In *Coming of Age: Scripps Institution of Oceaography: A Centennial Volume, 1903–2003*. Fisher, R.L., E.D. Goldberg and C.S. Cox, eds., 45–55. La Jolla: Scripps Institution of Oceanography, University of California, 2003.

Scripps Institution of Oceanography Timeline. 12 March 2008. <repositories.cdlib.org/ cgi/viewcontent.cgi?article=1031&context=sio/arch>

Shor, E.N. 1979. Tribute to Carl L.Hubbs (18 October, 1894 to 30 June, 1979). *Marine Biology* 55:81–2.

Shor, E.N., R.H. Rosenblatt, and J.D. Isaacs. Carl Leavitt Hubbs, October 18, 1894–June 30, 1979. Scripps Institution of Oceanography Archives. <http://repositories. cdlib.org/sio/arch/biog/hubbs/>

Steinbeck, Ricketts, and Holism in San Diego Bay

Ricketts. E.F. and J. Steinbeck (Preface). *Between Pacific Tides*. (Fifth edition). California: Stanford University Press, 1992.

Ritter, W.E. *The Unity of the Organism, or the Organismal Conception of Life*. Boston: Gorham Press, 1919.

Steinbeck, J., E. Ricketts, and R. Astro (Introduction). (Reprint edition). *The Log from the Sea of Cortez*. New York: Penguin Books, 1995.

Wilson, E.O. *Biophilia*. Cambridge, MA: Harvard University Press, 1986.

—. *Personal Interview*. October 2008.

Photo Credits*

Biodiversity
p. 13: Excurra interview—provided by Exequiel Excurra

Wetlands
p. 24: Peugh interview—provided by Jim Peugh

Shifting Baselines
p. 28 (left): 1936 tuna fishermen—Guy Bruni; (right): 2006 fishermen—Brad Fisher
p. 30: lenticulars—Randy Olson, Shiftingbaselines.org
p. 31: (left) old San Diego dock—San Diego Historical Society
p. 34: white seabass fisherman—Bob Hetzler

Green Sea Turtle
National Marine Fisheries Service Research Permit #1591

Peregrine Falcon
pp. 144, 146, 154 and 155—Will Sooter

Gray Whale
p. 179—Visual Services, SeaWorld San Diego

Ship Traffic
p. 197: Merk interview—Port of San Diego

Invasive Species
p. 204: *Caulerpa*—City of Carlsbad, CA
p. 207: Gonzalez interview—Marco Gonzalez

Climate Change
p. 238: Paul Neiman, NOAA
p. 243: Oppenheimer interview—the Trustees of Princeton University

Story of the Baiji River Dolphin
p. 248: Stephen Leatherwood/Press Association
p. 249: ©2006 baiji.org Foundation
p. 250: Yangtze River—http://en.wikipedia.org/wiki/Image:Yangtze-Ships.JPG;
p. 251: QiQi—Research Centre for Aquatic Biodiversity and Resource Conservation of the Chinese Academy of Sciences

Endangered Species Act
p. 262: Muir interview—provided by Rachel Muir

Stewards of the Bay
p. 275: Reznik interview—San Diego Coastkeeper

Sustainable Fishing
p. 290: old tuna fishermen—Guy Bruni

Carl Hubbs
p. 314: Laura Clark and Carl Leavitt Hubbs—Hubbs-SeaWorld Research Institute

*Unless listed here, all book photos were taken by students and teachers of High Tech High.

Graphic and Illustration Credits

Biodiversity
p. 5: map—Alex Bozzette

Methods
p. 41—Kelsey Hoffman

Abalone
p. 48—Rachel Bouffard

Green Sea Turtle
p. 54—Rachel Bouffard

California Brown Pelican
p. 64—Rachel Bouffard

White Pelican
p. 70—Rachel Bouffard
p. 77: DDT—Rachel Bouffard

California Least Tern
pp. 78, 80—Rachel Bouffard
p. 89—Kelsey Hoffman

Elegant Tern
p. 91—Rachel Bouffard

Gull-Billed Tern
p. 98—Kelsey Hoffman

Light-Footed Clapper Rail
p. 104—Kelsey Hoffman
p. 109—courtesy Richard Zemball

Black Oystercatcher
p. 112—Rachel Bouffard

Western Snowy Plover
p. 118—Rachel Bouffard
p. 121: population graph—Chris Nho

Black Skimmer
p. 124—Rachel Bouffard

Long-Billed Curlew
p. 131—Rachel Bouffard

Osprey
p. 136—Kelsey Hoffman

Peregrine Falcon
p. 145—Kelsey Hoffman

Northern Harrier
p. 158—Kelsey Hoffman

Burrowing Owl
p. 164—Rachel Bouffard

Gray Whale
p. 170—Rachel Bouffard
p. 171: migration map—Megan Morikawa

Sea Otter
p. 183—Rachel Bouffard
p. 186—Thomas Fernandez

Dredging
p. 212—Sean Curtice
p. 215: map—Integrated Natural Resources Management Plan

Endangered Species Act
p. 258—Thomas Fernandez

Stewards of the Bay
p. 266: logo—courtesy Port of San Diego
p. 268: logo—courtesy Wildcoast
p. 269: logo—courtesy Chula Vista Nature Center
p. 270: logo—courtesy California Coastal Commission

Marine Protected Areas
p. 303: map—courtesy World Wildlife Fund